Ultra-Narrowband Multispectral Imaging

This book provides insight into an unconventional modality of imaging where several spectral images are captured by a single snapshot under multilaser illumination, ensuring high-speed imaging within extremely narrow spectral bands. This method has three distinct advantages, if compared to common commercial multispectral imaging systems – considerably improved spectral selectivity (or color sensitivity) of imaging, avoided motion artifacts in the spectral image sets, and simpler/faster image processing.

This will be a valuable reference for laser and imaging professionals; photonics researchers and engineers; clinicians (e.g., dermatologists, plastic surgeons, oncologists); forensic experts; and students of physics, chemistry, biology, medicine, and engineering.

Key Features:

- Reviews techniques and applications of narrowband spectral imaging using multilaser illumination.

- Presents ten different prototypes for implementing the multi-spectral-line imaging method.

- Discusses applications of spectral line imaging for human skin diagnostics and forgery detection.

Ultra-Narrowband
Multispectral Imaging
Techniques and Applications

Janis Spigulis

CRC Press
Taylor & Francis Group
Boca Raton London New York

CRC Press is an imprint of the
Taylor & Francis Group, an **informa** business

First edition published 2025
by CRC Press
2385 NW Executive Center Drive, Suite 320, Boca Raton FL 33431

and by CRC Press
4 Park Square, Milton Park, Abingdon, Oxon, OX14 4RN

CRC Press is an imprint of Taylor & Francis Group, LLC

ISBN: 9781032757292 (hbk)
ISBN: 9781032760261 (pbk)
ISBN: 9781003476702 (ebk)

DOI: 10.1201/9781003476702

Typeset in Minion
by codeMantra

Contents

Author

Janis Spigulis graduated as physicist from the University of Latvia (UL) in 1973. He defended his PhD thesis on the collisional energy transfer in metal vapor mixtures in 1979 and obtained the Dr. Habil. Phys. degree in 1993 for research on fiber optics and optoelectronics. In 1997, Spigulis established and still leads the Biophotonics Laboratory at the Institute of Atomic Physics and Spectroscopy, UL; over the period 2004–2012, he was director of this institute. He has been promoted as Professor (1998), Fellow of SPIE (2011), academician of Latvian Academy of Sciences (LAS, 2012), and recipient of the Grand Medal of LAS (2021). Recent work of Prof. Spigulis aims at developing novel non-invasive optical methods and devices for applications in healthcare and other areas. He has supervised ~40 research projects with results published in >200 scientific papers and protected by >30 patents. Prof. Spigulis has been invited as keynote lecturer, committee member, or session chairman to tens of international conferences.

Abbreviations

3D	three-dimensional
ADC	amplitude-to-digital convertor
CMOS	complementary metal-oxide semiconductor microchip technology
FOV	field of view
FWHM	full width at half-maximum
HSI	hyperspectral imaging
LED	light-emitting diode
MSI	multispectral imaging
NIR	near infrared spectral range
PCB	printed circuit board
RGB	red, green, and blue color space
SMA	optical fiber connector standard
SMSLI	snapshot multi-spectral-line imaging
UV	ultraviolet spectral range

Introduction

"A PICTURE IS EQUIVALENT TO thousands of words" – probably you have heard this famous saying. Indeed, millions of pixels forming today's digital color photos contain plenty of information. Tremendous progress in the development of minia-ture electronic image sensors and sophisticated micro-lenses has made many of us "professional photographers," able to produce hundreds of high-quality photos or videos with our own mobile phone cameras. One may ask – can we get anything more than nice-looking pictures out of the digital image sensors? The answer certainly is YES, as the potential of digital imaging is far from being fully explored, and new imaging modalities are continu-ously developed and applied.

There are four main types of digital images – binary (black-and-white), grayscale, color, and multispectral images. This book gives insight into an unconventional modality of multispectral imaging within extremely narrow spectral bands, implemented by a single snapshot under specific multilaser illu-mination. Three main advantages, if compared to the common commercial multispectral imaging systems, are: (i) considerably improved spectral selectivity (or color sensitivity) of imaging, (ii) avoided motion artifacts in the spectral image sets, and (iii) sim-pler/faster image processing as integrals over the spectral bands of imaging are replaced by numbers of the fixed working wave-lengths. The basic principles of this approach and its progress achieved at the author's lab over the recent decade are reviewed

DOI: 10.1201/9781003476702-1

here, focusing on applications for non-contact human skin diagnostics and paper-based forgery detection.

The method itself was proposed and patented back in 2012; however, its implementation in functioning prototype devices and validation in real-life conditions became possible only after solving specific technical issues – first of all, ensuring uniform target illumination simultaneously by several laser lines, avoiding the laser speckle effects leading to multiple "grains" in the spectral line images, and experimental determination of relative spectral sensitivities of the imaging system at the working laser wavelengths. Designs of ten lab-developed prototypes implementing this method are described in the book, along with results of their laboratory, clinical, and/or forensic tests. The gained knowledge and experience facilitate the appearance of new equipment and protocols for better skin diagnostics and new modalities for highly sensitive detection of money and document forgeries, so contributing to improved public healthcare and security. Other interesting applications of this relatively simple but efficient technology may arise in the coming years, as well.

The presented material is structured into six chapters. Chapter 1 briefly explains the basics of spectral imaging, including the main principles of multispectral and hyperspectral imaging. Chapter 2 introduces the proposed snapshot multi-spectral-line imaging (SMSLI) method based on spectrally specific illumination. Chapter 3 describes the developed multilaser illumination designs, while Chapter 4 presents the main specifications of the lab-assembled prototype devices implementing such designs. Results of the test measurements confirming applicability of this method and the developed solutions for analysis and/or mapping of color pigments in clinical diagnostics and forgery detection are discussed in Chapters 5 and 6, respectively.

The author believes that the unconventional approaches and data presented here will be useful for professionals of imaging and laser technologies; photonics researchers and manufacturing engineers; clinicians (e.g., dermatologists, plastic surgeons, oncologists); forensic experts; students of physics, chemistry, biology, medicine, and engineering; and for anyone else interested in advanced imaging technologies.

CHAPTER 1

Basics of Spectral Imaging

W E CAN SEE DIFFERENT colors thanks to about 6 million tiny photoreceptor cells (called cones) in the retinal part of our eyes, functioning in line with ~120 million rod cells that ensure black-and-white vision. There are three main types of cones with different color sensitivities: S-cones (~10%) are more sensitive to blue light, M-cones (~30%) detect mainly the green region of the spectrum, while the L-cones (~60%) are responsible for the detection of red light. The proportion of signals arriving in the brain from numerous S-, M-, and L-cones determines the color we perceive at each point of the visual image. In brief, the color we see depends on the spectral responsivities of millions of micron-sized retinal photoreceptors.

Modern digital cameras, including those embedded in our mobile phones, mimic the anatomy of the human eye to produce electronic color images. Digital RGB (R – red, G – green, B – blue) image sensors, like eye retina, comprise millions of photosensitive cells, called pixels. To reproduce color, each light-responsive pixel contains three separate miniature photodetectors sensitive

DOI: 10.1201/9781003476702-2

to blue, green, or red light. The output signals from all pixels of the image sensor are further processed by an adapted software, resulting in a specific color/hue of every pixel of the image displayed on the color screen. The RGB model uses 8 bits each – from 0 to 23 -- for red, green, and blue colors. Each color also has values ranging from 0 to 255; this translates into millions of colors – 16,777,216 possible colors to be precise.

The number of pixels in digital cameras nowadays is reaching that of the human eye, so the performance of "electronic eye" in terms of spatial resolution is nearly the same. Concerning the color (or spectral) resolution, digital image sensors perform even better, thanks to their ability to separate the three main color components at every pixel. Our brain, in opposite, can only integrate simultaneously the inputs from all retinal cones into a specific color we perceive. The human eye is unable to gather spectral information at each point of the perceived image, while the data of an image captured by a digital RGB camera allows us to produce not only a color picture but also in parallel three separated color component images – red or R-image, green or G-image, and blue or B-image. Humans can see these three spectrally specific images only if using special color glasses – optical filters which selectively transmit just one of the three color bands. If we do so, some specific features in the field of view can be highlighted at each filtered spectral band, which allows analyzing the targeted objects remotely.

1.1 SPECTRAL REFLECTANCE

Colored items look so because they absorb some specific spectral components of the light shined on them, while the other spectral components are reflected. For instance, when sunlight (comprising all components of the visible spectrum) shines on a blue material, the green and red components of the incident light are absorbed; all green materials absorb the blue and red components of white light, and red materials absorb the blue and green components (Figure 1.1). Generally, every substance absorbs light only

FIGURE 1.1   Reflection of white light by blue, green, and red materials. Adapted from: https://www.shutterstock.com/search/color-absorption (royalty-free).

within specific spectral ranges, which depend on the chemical composition of this substance – every atom and every molecule, depending on the individual internal structure, has its own unique composition of the absorbed wavelength bands which compose its **absorption spectrum**. Material reflects only those spectral components which are not (or are minimally) absorbed, so the spectral distribution of the reflected light intensity – called **reflection spectrum** – bears rich information about the internal structure of the reflecting substance. To characterize light reflection quantitatively, the reflection coefficient – ratio of the reflected and incident light intensities, also named **reflectance** – is used. For layered and finite media, the term "reflectance" may be replaced by "**reflectivity**." Reflectivity is a measure of the intrinsic reflectance of a material, whereas reflectance may also depend on additional factors – like the material thickness and any materials of other structures, potentially present behind the surface material. When reflection occurs from thin layers of material, the internal reflection effects can cause the reflectance to vary with surface thickness. Reflectivity as a property of the material itself is the limiting value of reflectance when the sample becomes sufficiently thick; there is no difference between reflectivity and reflectance if light is reflected from an "infinitely thick" homogenous material.

If focusing only on one specific wavelength of light, the term "**spectral reflectance**" is used. It represents the ratio of the reflected light intensity I_r at the chosen light wavelength λ to that of the incident light I_o, or, mathematically, $R_\lambda = I_r(\lambda)/I_o(\lambda)$. More generally, this also relates to a small wavelength interval $\Delta\lambda$ centered at λ, if the reflectance within $\Delta\lambda$ does not change. Spectral reflectance is sometimes also named "monochromatic reflectance"; it may be specular or diffuse (if light can penetrate below the surface and be partly returned after multiple scattering). A reflection spectrum, also called a spectral reflectance curve, represents the changes of spectral reflectance over a selected wavelength range.

The chemical content of reflecting surfaces in many cases is non-uniform, i.e., there can be different spectral reflectance values at different spots of the surface to be examined – for example, of an art painting. In such cases, the x-y distribution maps representing the R_λ values at all points/pixels of the surface may provide valuable information about the object. Such distribution maps of R_λ are called **spectral images**. They can be obtained in a point-by-point manner (e.g., by means of a guided optical fiber) or, more commonly, all target pixels are captured at once by means of digital imaging cameras.

1.2 SPECTRAL IMAGING

Being related to only one narrow spectral interval, the spectral image contains specific information about the reflecting materials as each of them differently absorbs radiation of this selected wavelength or spectral band. Generally, **spectral imaging** refers to a group of analytical techniques that collect spectroscopic information and imaging information at the same time. The spectroscopic information tells us about the chemical makeup at the individual pixels of the image, allowing the absorbing pigment (chromophore) map of the imaged area to be produced. In parallel to "spectral imaging", the terms "spectroscopic imaging" and "chemical imaging" are also used.

There are two main approaches for obtaining spectral images – by narrowband spectral filtering of the image sensor at broadband (usually white) illumination of the target, or by means of spectrally narrowband illumination. The first approach is implemented using sets of spectral band filters, tunable acousto-optical filters, liquid crystal filters, or other tools for optical filtering. The second approach is based on the fact that all reflected light under quasi-monochromatic illumination (e.g., by a narrowband LED) relates only to this specific wavelength band. Spectral imaging then can be done by an unfiltered camera since only the spectral band of lighting determines the wavelength range related to the taken image.

The wavelength bands exploited for spectral imaging usually are much narrower than RGB bands of the consumer photo-cameras; the typical full width at half maximum (FWHM) is on the order of tens of nanometers. While the digital color cameras capture light across the three specially filtered RGB wavelength bands in the visible spectrum, the spectral imagers may operate in a much wider spectral range, spanning from ultraviolet to infrared regions – it mainly depends on the spectral sensitivity of the exploited image sensor. The silicon-based matrix photodetectors are sensitive in the 300–1,100 nm range; if unfiltered, they can capture images also in the UV (300–400 nm) and NIR (700–1,100 nm) spectral ranges. InGaAs image sensors bridge the gap between NIR wavelengths in the 950–1,700 nm range.

1.3 MULTISPECTRAL IMAGING

Methods are called *multispectral imaging* (abbreviated **MSI**) when they provide images of the same target within two or more different wavelength ranges. The number of working spectral bands for MSI typically varies between 3 and 10; they usually do not overlap, or their spectral overlap is insignificant. The concept of multispectral imaging is illustrated in Figure 1.2. The wavelength bands selected for imaging may differ in terms of their number (5 in this case), spectral width (FWHM), and the achieved sensitivity; often in between the working channels, there are spectral regions with negligible responsivity of the target. The imaged spectral ranges can be customized for a particular application – for example, remote plant monitoring imagers

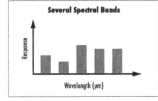

FIGURE 1.2 Conceptual illustration of multispectral imaging within five separated spectral bands. With permission, adapted from https://www.edmundoptics.com/knowledge-center/application-notes/imaging/hyperspectral-and-multispectral-imaging/.

usually include the chlorophyll absorption bands 420–480 nm and/or 630–690 nm. The MSI working bands may be extended to the invisible parts of the spectrum which allows extracting additional information beyond the human vision abilities.

Multispectral imaging can be implemented in several ways. For example, the conventional camera can be supplemented by a rotating wheel containing multiple optical bandpass filters for sequential selection of specific wavelength regions. The filters can be turned in front of the camera lens (including smartphone cameras, Figure 1.3c) both manually or automatically, e.g., using a step motor. No matter how, a set of spectral images of the target related to all filter-transmitted spectral bands is obtained. Use of a rotating filter wheel is a relatively simple and robust MSI technique;

 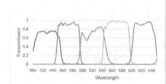 

a   b   c

FIGURE 1.3 Rotating filter wheel (a), the potential MSI working wavelength bands (b), and multispectral smartphone with a filter wheel in front of camera lens (c). (a) Copyright Company Seven 2003, used by permission; (b) Open access source https://library.imaging.org/ei/articles/30/16/art00006; (c) With permission from https://hackaday.io/project/185259-multispectral-imaging-smartphone-camera.

still, issues concerning the spectral selectivity, dimensions of the setup, vibrations, synchronization of exposures, and time for collecting the spectral image set may cause some problems during specific applications.

One can also employ a stationary combination of several cameras, each of them equipped with a fixed bandpass filter for limiting the camera sensitivity to a certain spectral region. This design notably speeds up the capturing process of the multispectral image set but makes the MSI equipment much more expensive; alignment of several cameras to have precisely the same field of view (FOV) is an additional challenge.

Fast single-snapshot multispectral imaging is also possible using only one camera comprising a multiband image sensor. The standard RGB image sensors ensure capturing of three overlapping broadband spectral images by a single snapshot. Recently, also imaging within more than three spectral bands by so-called mosaic image sensors has become possible. Thanks to specific multi-layer structure, every pixel of such sensors enables $n > 3$ output channels, each related to a different wavelength band.

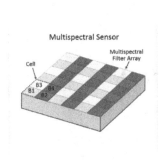
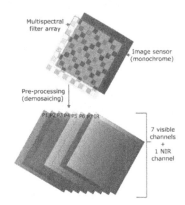

FIGURE 1.4 Design principles of 4-band (left) and 8-band (right) mosaic multispectral image sensors. With permission from https://spectraldevices.com/pages/technology and from the open access source https://www.mdpi.com/1424-8220/14/11/21626.

For example, Figure 1.4 illustrates the designs of mosaic multispectral image sensors for snapshot recording of four and eight spectral images. Spectral images in such cases are extracted from the single digital image data set by an adapted software, as it is done in the RGB cameras. Snapshot MSI cameras with multiband image sensors ensuring a larger number of spectral channels (e.g., 16, 32) are also available. However, the filter-selected spectral bands of such image sensors usually overlap - this limits the spectral selectivity of imaging. Besides, so far the mosaic sensor cameras are too expensive for wide routine applications.

As mentioned above, one more option for obtaining spectral images is their capturing under spectrally specific narrowband illumination. Inexpensive broadband (e.g., black-and-white) cameras suit well for this MSI modality as the spectral band of imaging depends only on that of illumination. Sequential capturing of images, each of them related to a different spectral band, is ensured by re-switching the narrowband illumination sources, e.g., a set of color LEDs or some appropriately filtered broadband light source. This approach is efficient if the distance between the

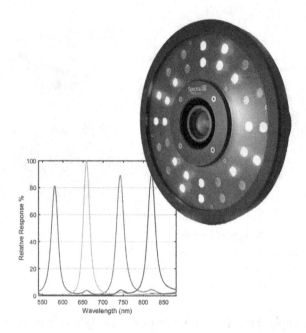

FIGURE 1.5 Multispectral LED illuminator ring around the camera lens. With permission from https://spectraldevices.com/products/multispectral-imaging-system-for-agriculture-1.

camera and target is relatively short so the ambient illumination can be avoided by screening the target area. Various multi-LED illumination designs have been developed for this application, including an LED ring that surrounds the camera lens (Figure 1.5). Several LEDs emitting the same spectral band are located uniformly around the lens and are switched on for capturing the image within this single spectral range. After switching them off, another set of LEDs emitting in some other wavelength band is switched on to capture the next spectral image. This way a set of spectral images related to all LED-emitted wavelength bands is collected.

MSI technology was originally developed for reconnaissance and military target identification. Early space platforms incorporated filter-based multispectral imagers to map details of the Earth surface related to coastal boundaries, vegetation, and landforms. Numerous multispectral imaging cameras are now installed on space satellites, airplanes, and drones. There are also multispectral imagers embedded in industrial settings, as well as a wide variety of hand-held MSI devices for different mobile applications.

1.4 HYPERSPECTRAL IMAGING

The term **hyperspectral imaging** (abbreviated **HSI**) is applied in cases when numerous spectral images related to adjacent/ overlapping wavelength bands are taken, so that a full optical spectrum for each pixel of the image can be reconstructed. The number of spectral images collected by HSI technology typically exceeds 10 and may reach even hundreds of spectral channels for the same target image. This provides much more data than multispectral imaging, allowing more specific analysis of materials and substances. The full HSI dataset forms the so-called hyperspectral image cube with two spatial dimensions and one spectral dimension – see Figure 1.6 for illustration.

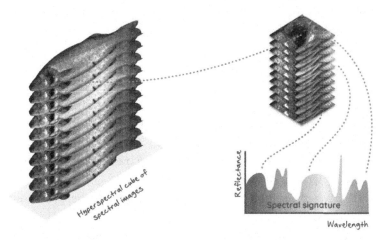

FIGURE 1.6 Conceptual illustration of hyperspectral imaging with extracted reflectance spectrum for a selected pixel of the taken image. With permission from https://condifood.com/hyperspectral-imaging/.

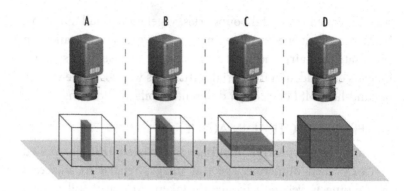

FIGURE 1.7 The four primary methods for hyperspectral data acquisition: A – point scanning or whiskbroom mode, B – line scanning or pushbroom mode, C – spectral scanning or area scanning mode, D – single snapshot mode. With permission from https://www.edmundoptics.eu/knowledge-center/application-notes/imaging/hyperspectral-and-multispectral-imaging/.

There are four primary methods used for hyperspectral image acquisition (Figure 1.7). The whiskbroom method is a point scanning process that acquires the spectral information for one spatial coordinate at a time. The pushbroom method is a line scanning process where a linear spatial movement is required as a row of pixels scans over an area to capture the spectral and positional information. The plane scanning method images the entire 2D area at once, but at each of the adjacent wavelength intervals (i.e., collects a set of spectral images). The fourth mode of hyperspectral image acquisition, snapshot HSI (e.g., by a 32-band camera), enables collection of the whole hyperspectral data set within a single camera exposition.

HSI cubes contain a huge amount of data to be processed and stored. For example, one can consider hyperspectral imaging at 50 spectral channels with a spatial image resolution of $1,024 \times 1,024$ pixels. Then each hyperspectral image set contains over 50 million intensity values, each one with, e.g., 16 bits, which makes nearly a Gigabit data set per single HSI cube. Consequently, the HSI equipment should involve not only advanced devices for acquisition of numerous spectral images but also sophisticated software for extracting the desired spectro-spatial information. This makes the HSI implementation rather complicated and much more expensive if compared with MSI. The non-snapshot HSI procedures also take more time (typically from tens of seconds to minutes) and therefore motion artifacts in spectral images may appear. A specific aspect of HSI is utility – how efficiently the time and computing resources are spent. There are many applications where only a limited number of spectral channels provide useful information about the targeted object, and capturing/processing the other tens or hundreds of spectral images is just a waste of time and resources; in such cases, MSI technique is clearly preferable.

Hyperspectral imaging is widely applied in industry and healthcare. For example, airborne and space-based spectral imaging instruments are used for geological surveys, e.g., for finding ores or oil reservoirs, as well as for environmental monitoring. Detailed spectroscopic image information can help to ensure food safety and to support doctors in medical diagnostics. Other applications include industrial processing, forensic science, biomedical research, investigation of artworks, and many more.

This chapter provided only a brief insight into the basic aspects of spectral imaging and its modalities. More comprehensive data and details are presented in specialized books and review articles; just a few of recommended ones are [1–5].

Snapshot Multi-Spectral-Line Imaging

A S SHOWN IN THE previous chapter, multispectral imaging is an efficient tool for remote characterizing of the target object's properties by revealing specific spectral reflectance's from different spots of its surface. The performance of this technique critically depends not only on a properly chosen central (peak) wavelengths of the working bands, but also on their spectral bandwidth (FWHM) and how well the bands are separated – whether the neighboring spectral bands overlap or not. Ideally, the imaging spectral bands should be fully separated and as narrow as possible. High spectral purity, also called spectral selectivity of imaging and/or color sensitivity, is an important issue for many applications of MSI.

Imagine you are sitting in a room illuminated by some laser system emitting just one laser wavelength (also called laser line), for instance, only the red wavelength 650 nm. If so, everything you see around are spectral images of the single wavelength 650 nm,

DOI: 10.1201/9781003476702-3

as there are no other spectral components to be reflected. Any camera in such conditions would capture exclusively the 650 nm spectral line images, as well. What is most important – the spectral band of imaging would be extremely narrow as the laser emission linewidth (FWHM) typically is below 0.1 nm and can be reduced to 10^{-5} nm or even less, if special measures are undertaken [6]. Combining several laser spectral line images taken at different illumination wavelengths may provide record-high spectral selectivity (correspondingly, color sensitivity) of imaging, so opening new prospects for high-performance MSI.

Another important MSI performance criterion is the time needed for acquisition of the spectral image set. It determines not only the ease of operation but also the imaging quality, especially if the target object or MSI device can move. In such cases, the position of the object in different spectral images may be shifted, which causes the blurring effect and significantly complicates the multispectral image processing procedure. To avoid this, the spectral image set should be collected as fast as possible, preferably by a single snapshot.

Both mentioned performance challenges can be met if the snapshot multi-spectral-line imaging (abbreviated SMSLI), a method developed in the author's lab over the recent decade, is used. This chapter gives insight into the main concepts and options of SMSLI.

2.1 SNAPSHOT DUAL SPECTRAL LINE IMAGING

One of the MSI application areas is color pigment analysis and mapping of their distribution over the target area. In dermatology, distribution maps of skin hemoglobin – a component of red blood cells, strongly absorbing in the blue-green region of the spectrum and ~100 times weaker in the red region [7] – help to diagnose skin diseases. Our study [8] demonstrated reliable 2D-mapping of skin hemoglobin changes by means of a hyperspectral camera in the wavelength range 500–700 nm. Trying to save resources and simplify the procedure, the potential of dual spectral line imaging for *in vivo* skin hemoglobin mapping was studied in [9]. This research was aimed to find out if the spectral line images at only two laser wavelengths – green 532 nm (high absorption) and red 635 nm (low absorption) – could provide physiologically feasible description of skin blood supply changes after occlusion of blood vessels. Parallel measurements were taken by a 51-channel hyperspectral imaging system, a much more complicated and resource consuming equipment. Illumination of skin was provided simultaneously by two cw lasers (532 and 635 nm) operating in 1…10 mW power range. Radiation of both lasers via a Y-shaped optical fiber assembly formed a joint beam that was expanded to illuminate ~5 cm spot on the skin. Initial spectral reflectance measurements indicated that skin fluorescence under double-laser illumination could be neglected. Skin spectral line images were captured by a snapshot of RGB camera with signal output to PC. More details on the measurement setup are provided in Paragraph 3.1.

The reflected 532 nm light mostly was detected by the CMOS sensor green (G) sub-pixels, and the 635 nm light – mostly by the red (R) subpixels (Figure 2.1). In this situation, the consumer RGB camera worked as a 2-channel multi-spectral detector acquiring both spectral line images at once, by a single snapshot.

As shown in Figure 2.1, the two sensitivity bands of the image sensor are partly overlapping, therefore the detected green light at

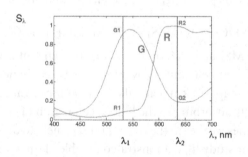

FIGURE 2.1 Spectral sensitivity G- and R-curves of the used RGB detector (manufacturer's data) and the specific sensitivities at two laser illumination wavelengths [10].

532 nm also slightly contributes to the R-band output (point R1), and the detected red light at 635 nm also contributes to the G-band output (point G2). Even assuming that the G-band output of RGB camera was fully determined by reflection of the 532 nm laser line and that of the R-band – by reflection of the 635 nm laser line, physiologically feasible results in this experiment were obtained [9]. Moreover, exact corrections related to the crosstalk between the G-band and R-band outputs are possible [10]. We suppose that (i) the RGB spectral sensitivity curves are known – given by the manufacturer and/or have been previously measured, (ii) they are uniform – the same for all pixels over the image sensor, (iii) linear photovoltaic responses at both G- and R-channels are ensured, and (iv) the target is not fluorescing, or its fluorescence intensity is negligibly low. If so, the detected digital signal values at the G- and R-channels are proportional to the respective spectral sensitivities at both wavelengths, marked as $S(G1)$, $S(G2)$ and $S(R1)$, $S(R2)$; then the output signal ratios $S_G = S(G1)/S(G2)$ and $S_R = S(R1)/(R2)$ are fixed values.

If the image of a white reflector (e.g., white paper) at bichromatic illumination is analyzed, the signals detected at both spectral channels can be expressed as:

$$G = G1 + G2, \qquad (2.1)$$

$$R = R1 + R2, \qquad (2.2)$$

$$G1 = G2 \cdot S_G, \qquad (2.3)$$

$$R1 = R2 \cdot S_R. \qquad (2.4)$$

There are four known/measured values (G, R, S_G, and S_R) and four unknown values that can be easily found by solving the system (equations (2.1)–(2.4)):

$$G1 = G \cdot S_G/(1 + S_G), \qquad (2.5)$$

$$G2 = G/(1 + S_G), \qquad (2.6)$$

$$R1 = R. \, S_R / (1 + S_R), \tag{2.7}$$

$$R2 = R/(1 + S_R). \tag{2.8}$$

Consequently, exact crosstalk corrections of the R- and G-values at bi-chromatic illumination can be applied. For example, to find the "true" value $G1$ responsible for the signal at λ_1 that has been detected in the G-channel, simple extraction according to equation (2.1) should be applied:

$$G1 = G - G / (1 + S_G) = G \, [1 - 1 / (1 + S_G)]. \tag{2.9}$$

The same considerations remain valid for any of the two other RGB band combinations (R-B and G-B), concluding that the impact of crosstalk can be fully considered in any situation of RGB imaging under double spectral line illumination. In such circumstances, two spectral images, each related to an extremely narrow spectral band (equal to the illumination laser linewidth), can be successfully extracted from the RGB image dataset.

To calculate the spectral images of the target, let us assume k_1 to be the spectral reflectance at the wavelength λ_1 and k_2 the spectral reflectance at the wavelength λ_2. Both k_1 and k_2 will influence the values of green and red output signals G' and R' at each pixel of the target image:

$$G' = k_1 \cdot G_1 + k_2 \cdot G_2 = \frac{k_1 G S_G}{(1 + S_G)} + \frac{k_2 G}{(1 + S_G)}, \tag{2.10}$$

$$R' = k_1 \cdot R_1 + k_2 \cdot R_2 = \frac{k_1 R S_R}{(1 + S_R)} + \frac{k_2 R}{(1 + S_R)}. \tag{2.11}$$

In result, crosstalk-corrected spectral reflectance values k_1 and k_2 at each pixel or pixel group of the target image can be obtained as

$$k_1 = \frac{G'(1 + S_G)R - R'(1 + S_R)G}{GR(S_G - S_R)}, \tag{2.12}$$

$$k_2 = \frac{GS_G R'(1+S_R) - RS_R G'(1+S_G)}{GR(S_G - S_R)}. \qquad (2.13)$$

The relative spectral reflectance at each spot (pixel) of the target can be easily calculated, as well:

$$k = \frac{k_2}{k_1} = \frac{GS_G R'(1+S_R) - RS_R G'(1+S_G)}{G'(1+S_G)R - R'(1+S_R)G}. \qquad (2.14)$$

Expression (2.14) is useful if the object's parameter to be mapped depends on the relative spectral reflectance at two specified wavelengths – e.g., in the case of color pigment mapping.

Still, the question about reliability of the available RGB sensitivity curves remains. The matrix image sensors like other semiconductor components are manufactured under a specific technology, and the mass-produced sensors made under the same technology do not always exhibit exactly the same signal sensitivity parameters. That is the reason why the suppliers of image sensors and RGB cameras usually do not provide the individual RGB spectral sensitivity curves. Instead, the customer gets acquainted with "typical" spectral sensitivities related to a specified series of products, with a comment that for some products of the same series they could be different. Therefore, it is highly recommended to validate the manufacturer's data by independent measurements.

One of the options is to capture white reflector images under monochromatic (single wavelength) or narrowband illumination at a wider spectral range. Such a procedure was described in paper [10] with respect to the used RGB camera. It was performed by comparing the measured RGB output signal ratios at several narrow wavelength bands with those calculated from the manufacturer's data (Figure 2.1). The measurements were taken by capturing white paper images under monochromatic (lasers) or quasi-monochromatic (narrowband LEDs) illumination. The expected ratios (i.e. those calculated from the manufacturer's provided B- and R-channel spectral sensitivity curves at the working wavelengths) were compared to the measured ratios at these wavelengths. The obtained correlation was good ($R^2 = 0.9735$); the main disagreement was observed in cases of laser illumination (405, 532 nm) which may be explained by slightly uneven surface illumination due to laser speckles – the "grains" or brighter spots caused by the constructive interference of laser light, where local nonlinearity of photo-response could take place.

To summarize, our early studies confirmed the possibility of snapshot spectral imaging within two ultra-narrow spectral ranges, determined by the illuminating laser's spectral linewidths (which are typically less than 0.1 nm). This technique was further expanded for three, four, and more spectral line imaging.

2.2 SNAPSHOT TRIPLE SPECTRAL LINE IMAGING

The above-described approach can be further extended to the situation when the target area is illuminated simultaneously by three spectral lines (Figure 2.2). It is supposed that the selected three wavelengths fit within the RGB spectral sensitivity interval and the intensities of all three spectral lines are equal; linearity of the image sensor photo-response is a condition [11].

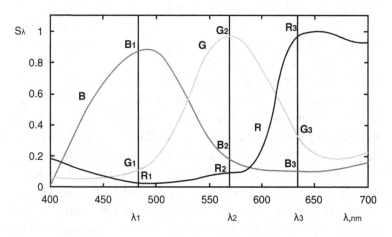

FIGURE 2.2 Spectral sensitivity B-, G-, and R-curves of the RGB detector and the specific sensitivities at three laser illumination wavelengths [11].

Each of the color channels (R, G, and B) then detects a certain part of the signal at all three illumination wavelengths:

$$R = R_1 + R_2 + R_3, \qquad (2.15)$$

$$G = G_1 + G_2 + G_3, \qquad (2.16)$$

$$B = B_1 + B_2 + B_3, \qquad (2.17)$$

where R_1, G_1, and B_1 are the signals registered in the R-, G-, and B-channels at wavelength λ_1; R_2, G_2, and B_2 are the signals registered in the R-, G-, and B-channels at the second wavelength λ_2; and R_3, G_3, and B_3 are the signals registered in the R-, G-, and B-channels at the third wavelength λ_3. The ratios of spectral sensitivities of the R, G, and B bands of the image sensor (and, consequently, the ratios of the digital signal values) again are fixed numbers: $S_{R12} = \dfrac{S(R_1)}{S(R_2)}$,

$S_{R13} = \dfrac{S(R_1)}{S(R3)}$, $\quad S_{R23} = \dfrac{S(R_2)}{S(R_3)}$, $\quad S_{G12} = \dfrac{S(G_1)}{S(G_2)}$, $\quad S_{G13} = \dfrac{S(G_1)}{S(G_3)}$,

$S_{G23} = \dfrac{S(G_2)}{S(G_3)}$, $S_{B12} = \dfrac{S(B_1)}{S(B_2)}$, $S_{B23} = \dfrac{S(B_2)}{S(B_3)}$. After appropriate substitutions, all RGB signal components can be expressed analytically as:

$$
\left\{
\begin{aligned}
R_1 &= \frac{R}{1 + \dfrac{1}{S_{R12}} + \dfrac{1}{S_{R13}}}, R_2 = \frac{R}{S_{R12} + 1 + \dfrac{1}{S_{R23}}}, R_3 = \frac{R}{S_{R13} + S_{R23} + 1}, \\[2ex]
G_1 &= \frac{G}{1 + \dfrac{1}{S_{G12}} + \dfrac{1}{S_{G13}}}, G_2 = \frac{G}{S_{G12} + 1 + \dfrac{1}{S_{G23}}}, G_3 = \frac{G}{S_{G13} + S_{G23} + 1}, \\[2ex]
B_1 &= \frac{B}{1 + \dfrac{1}{S_{B12}} + \dfrac{1}{S_{B13}}}, B_2 = \frac{B}{S_{B12} + 1 + \dfrac{1}{S_{B23}}}, B_3 = \frac{B}{S_{B13} + S_{B23} + 1}.
\end{aligned}
\right.
\qquad (2.18)
$$

This solution leads to the RGB crosstalk corrections of white-reflected signals at simultaneous three spectral line illumination. The corrected reflection signal value for a particular image pixel (or selected group of pixels) at the wavelength λ_1 will be the sum $(R_1 + G_1 + B_1)$, at λ_2 the sum $(R_2 + G_2 + B_2)$, and at λ_3 the sum $(R_3 + G_3 + B_3)$.

Spectral reflectances k_1, k_2, and k_3 at the three specified wavelengths can be mapped now with respect to the above-described crosstalk corrections. The sample-reflected R'', G'', and B'' values at the output of a particular image pixel (or pixel group) are:

$$\begin{cases} R'' = k_1 R_1 + k_2 R_2 + k_3 R_3 \\ G'' = k_1 G_1 + k_2 G_2 + k_3 G_3. \\ B'' = k_1 B_1 + k_2 B_2 + k_3 B_3 \end{cases} \quad (2.19)$$

After transformations, the following expressions for the k_1, k_2, and k_3 values at each image pixel are obtained:

$$k_1 = \frac{B''G_3 R_2 - B_3 G'' R_2 - B'' G_2 R_3 + B_2 G'' R_3 + B_3 G_2 R'' - B_2 G_3 R''}{B_3 G_2 R_1 - B_2 G_3 R_1 - B_3 G_1 R_2 + B_1 G_3 R_2 + B_2 G_1 R_3 - B_1 G_2 R_3},$$

$$(2.20)$$

$$k_2 = \frac{-B''G_3 R_1 + B_3 G'' R_1 + B'' G_1 R_3 - B_1 G'' R_3 - B_3 G_1 R'' + B_1 G_3 R''}{B_3 G_2 R_1 - B_2 G_3 R_1 - B_3 G_1 R_2 + B_1 G_3 R_2 + B_2 G_1 R_3 - B_1 G_2 R_3},$$

$$(2.21)$$

$$k_3 = \frac{B''G_2 R_1 - B_2 G'' R_1 - B'' G_1 R_2 + B_1 G'' R_2 + B_2 G_1 R'' - B_1 G_2 R''}{B_3 G_2 R_1 - B_2 G_3 R_1 - B_3 G_1 R_2 + B_1 G_3 R_2 + B_2 G_1 R_3 - B_1 G_2 R_3}.$$

$$(2.22)$$

An initial assumption was that the intensities of all three illumination spectral lines are equal. If it is not the case, the related relative spectral sensitivities should be corrected accordingly.

For example, if the intensity ratios of illuminating spectral lines I_1, I_2, and I_3 are 1.00/0.83/0.79, the correction coefficients 0.83 and 0.79 are to be applied for S_2 and S_3, respectively.

To summarize, parallel illumination of the target by three spectral lines and appropriate processing of the RGB image data theoretically provide the opportunity to calculate the spectral reflectance of the targeted area for each image pixel at each of the three wavelengths, with further obtaining of the corresponding three spectral line images. To check this statement, the above-described model was experimentally validated by RGB image measurements of a four-color target [11].

The measurement setup comprised an objective-supplied CMOS camera (*USB 2 UI-1226LE-C, IDS* with known RGB sensitivities) and three stabilized cw laser modules emitting at 405, 532, and 650 nm with output powers in the range 10…15 mW. An optical fiber assembly transformed the three laser beams into randomly distributed 21 beams from the fiber ends placed ring-wise around the camera objective. The four-color target area was located 5 cm from the CMOS objective lens. Laser output powers and the exposition time of CMOS sensor were carefully adjusted to provide linearity of photovoltaic responses at all spectral combinations used in experiments. All automatic features of the RGB sensor were switched off. More details on this setup are available in Paragraph 3.2.

Flat paper target composed of four colored segments (red, green, blue, and white – Figure 2.3a) was used. The colors were selected to provide a different spectral image at each illumination wavelength (Figure 2.3b). All three combinations of dual-wavelength illumination were tested, as well as the simultaneous triple-wavelength illumination. The formulas presented above were used to extract crosstalk-corrected spectral images from the recorded digital RGB data sets.

In the case of bi-chromatic 405 and 650 nm illumination, the R-channel output values related to the green and white sectors of the target remained practically unchanged, as well as the B-channel values related to the blue sector. The illumination combining 405

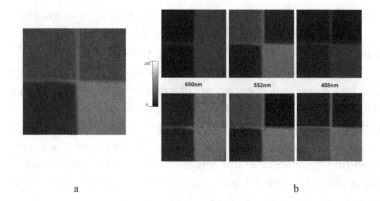

FIGURE 2.3 The four-color target design (a) and its spectral images, obtained under illumination of three laser spectral lines (b) [11].

and 532 nm wavelengths did not cause any real corrections to the G-channel output signal values related to the green and white sectors of the target, as well as to the B-channel output signal values related to the green and blue sectors; the green sector signals in both channels remained practically the same.

At a combined 532 and 650 nm illumination, the crosstalk-corrected target images detected at the R- and G-channels visually appeared different; those related to the red and green sectors of target experienced quite significant changes after corrections, while those related to the blue and white sectors changed insignificantly. In particular, in the red sector, R-value increased for 15% and G decreased for 53%; in the green sector, R decreased for 22% and G increased for 13% [11]. Consequently, the RGB crosstalk correction is a significant issue to be observed in the spectral line image processing procedure.

The crosstalk corrections at simultaneous three wavelength (405, 532, and 650 nm) illumination were significant in most combinations, except for the green sector at the R-channel and for the blue sector at the G-channel. Figure 2.4 shows that all three gray-scale spectral images after the crosstalk corrections have slightly changed.

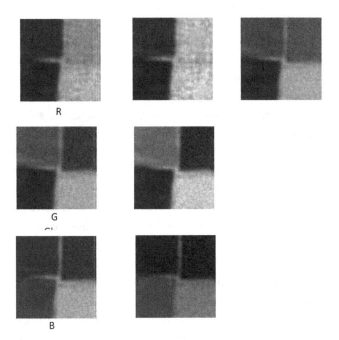

FIGURE 2.4   650, 532, and 405 nm spectral images under simultaneous triple wavelength illumination, detected at the R-, G-, and B-channels without (left) and with (right) crosstalk correction. Upper right: RGB image of the target at combined 650-532-405 nm illumination [11].

The presented spectral images were not calibrated in units of spectral reflectance. It can be done technically, but our primary goal was to illustrate how much the RGB crosstalk corrections influence the detector output data used for spectral image processing. In most cases, the variations were considerable.

To summarize, the above-presented results confirmed the technical possibility of transforming digital RGB cameras into inexpensive single snapshot ultra-narrowband multispectral imagers by adding an adjusted triple-wavelength illumination source and appropriate software. Comparative analysis of the R, G, and B band signals detected at the pixel's outputs allowed specifying much narrower wavelength intervals of the spectral images, compared to the *R*, *G*, or *B* spectral sensitivity bands of the conventional RGB image sensors or to the narrowband LED emission. Direct use of all registered *R*, *G*, and/or *B* pixel output signal values with subsequent corrections of the spectral RGB inter-channel crosstalk represents a fast, cost-efficient, and robust way to obtain high-performance spectral images in situations when spectrally specific target illumination is possible.

The general concept of snapshot triple spectral line imaging in comparison with the traditional spectral band imaging is illustrated in Figure 2.5. Three main advantages of the proposed approach are significantly improved spectral selectivity, faster acquisition of the spectral image set (single snapshot instead of sequential image capturing), and simpler image processing thanks to avoided integration over the spectral bands of imaging. This method and the associated device have been patented [12]. One more patented solution [13] relates to situations when the targets are relatively dark – to avoid disproportionately differing spectral reflectance values that lower the quality of spectral images, it was proposed in such cases to replace the white reference reflector with a calibrated gray reflector and to amend correspondingly the expressions (equations (2.18)–(2.22)).

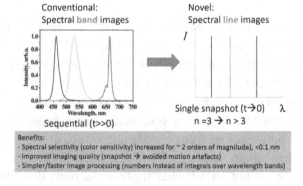

FIGURE 2.5 The snapshot triple spectral line imaging concept and its main advantages (author's drawing).

2.3 FAST ACQUISITION OF MORE THAN THREE SPECTRAL LINE IMAGES

One may ask if the SMSLI concept is limited only to three spectral line images, or it can be extended to a larger number of ultra-narrowband spectral images? Yes, it can be extended, and two of the possible implementation options are discussed below.

The first and most obvious option is to replace the common RGB image sensor with a more advanced mosaic sensor, capable of detecting images in 4, 8, 16, 32, or even a larger number of sensitivity bands. To essentially improve the spectral purity of multispectral imaging by means of such sensors, the corresponding number of ultra-narrowband spectral line images can be obtained at illumination comprising the same number of discrete spectral lines, each of them spectrally positioned within one of the sensor's sensitivity bands. The basic SMSLI concept remains the same, only equations (2.15)–(2.22) should be modified according to the new number of working wavelengths.

As the simplest case, let us consider snapshot four spectral line imaging. The device described in [14] comprised a four-band RGB-NIR camera (MSC-RGBN-1-A, *Spectral Devices Inc.*, CA) with spectral sensitivity curves shown in Figure 2.6. Four wavelengths illumination of the target area was performed using an RGB laser simultaneously emitting three spectral lines (450, 523, 638 nm) and a NIR laser emitting at 850 nm. As seen in Figure 2.6, each of the working wavelengths fits within one of the image sensor's sensitivity curves. This device ensured snapshot capturing of four spectral line images related to the above-mentioned wavelengths. More details on this device design can be found in Paragraph 4.4.

Even if only a three-band RGB image sensor is available, more than three spectral line images can be obtained in sub-second time intervals by a method proposed in [15]. It requires the illumination system capable of emitting $n > 3$ spectral lines, under the following conditions:

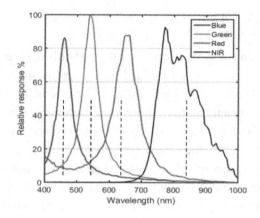

FIGURE 2.6 Spectral sensitivities of the four-band image sensor; dotted lines represent the four illumination wavelengths used for ultra-narrowband multispectral imaging [14].

The wavelengths of illuminating spectral lines are within the RGB camera sensitivity range and their intensities are equal or mutually comparable.

It is possible to quickly turn on and off the lighting by each spectral line or by pre-defined set(s) of spectral lines.

The object can be evenly illuminated by all available triple spectral line combinations, ensuring the linearity of photodetection,

RGB spectral sensitivities of the image sensor at all applied wavelengths are known.

First, an RGB image is captured under one combination of three spectral line illumination, with subsequent extraction of three spectral line images related to the illumination wavelengths. As the next step, a second RGB image is captured under a different combination of triple spectral line illumination, where the wavelengths of all spectral lines (or at least of one) are different.

This way one to three new spectral line images are obtained. If there is a need for a higher number of spectral line images, this procedure can be repeated using other combinations of illumination spectral line triplets. Some of the techniques suitable for quick replacement of the spectral line illumination sets are as follows:

programmed pulsed switching of the power supplies connected to illuminating lasers or other light sources emitting discrete spectral lines,

use of electro-optical switches, e.g., Pockels cells in front of the illuminating lasers or other spectral line emitters,

use of mechanically movable spectral filters, e.g., rotating interference filter wheels, in front of continuously emitting multi-spectral-line emitters, and

use of electronically controlled variable optical bandpass devices, e.g., liquid crystal filters in front of the multi-spectral-line emitters.

2.4 EXPERIMENTAL DETERMINATION OF CAMERA'S RELATIVE SENSITIVITIES AT SELECTED WAVELENGTHS

Following the considerations of Paragraph 2.2, the set of multi-spectral-line images can be obtained without knowledge of the photo-camera absolute sensitivities at the working wavelengths – it is enough to know their relative spectral sensitivities. Even having no idea about the spectral sensitivity curves of the used camera, the relative sensitivities at the selected wavelengths can be determined experimentally by a method patented in [16]. To determine the relative spectral sensitivities of a digital camera at selected wavelengths, a multi-input lightguide is used, with each input connected to a different spectral line emitter, e.g., laser. Each of them emits a single selected wavelength λ_i, and the radiation of each selected spectral line is sequentially turned on and off. At the joint output of the lightguide, the radiation power at every selected wavelength is measured by a laser power meter or a similar device. According to the measurement results, the relative radiation power P_i of each output spectral line is determined. As the next step, the output of the lightguide is directed toward a calibrated reference reflector, whose reflection coefficient R is constant in the selected spectral interval or has a known reflectivity for each of the used wavelength. Then the camera takes images of the reflector at each of the selected illumination wavelengths λ_i, and a group of pixels comprising a complete image of the illuminated reflector is determined in the captured images. Further, the average pixel's group output signal value A_{ij} is calculated for each of the selected illumination wavelengths λ_i at each of the camera spectral channels j. In total, $i \times j$ signal values are obtained. The relative spectral sensitivity $S_j(\lambda_i)$ for each individual camera's spectral channel j at each selected wavelength λ_i is determined using the formula

$$S_j(\lambda_i) = P_i \times A_{ij} \times R_i, \qquad (2.23)$$

where P_i denotes the measured relative power of the i-th spectral line output radiation, A_{ij} the mean value of the pixel group signals at the j-th photodetection band of the used camera, R_i the known reflection coefficient value of the target at wavelength λ_i. To determine the relative spectral sensitivity of the camera to radiation at one selected wavelength λ_i over all photodetection channels, all $S_j(\lambda_i)$ values should be summed.

Instead of several light sources, a single discrete line spectrum light source can be used at the input of the lightguide, with the possibility of filtering each selected spectral line – for example, a hollow cathode spectral lamp equipped with a set of appropriate interference filters.

The schematic of a device implementing this method is shown in Figure 2.7. It includes a lightguide 1 with several inputs connected to the selected spectral line emitters 2a, 2b, 2c,..., 2i, as well

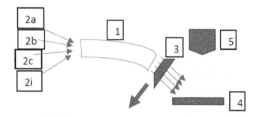

FIGURE 2.7   Scheme of the equipment for experimental determination of the photo-camera relative spectral sensitivities at the selected wavelengths. 1 – lightguide, 2 – spectral line sources, 3 – radiation power meter, 4 – calibrated reflector, 5 – photo-camera. The arrow denotes to the removal of the power meter after performing the lightguide output power measurements [16].

as a calibrated radiation power meter 3 at the lightguide output (to be removed after the power measurements), and a calibrated reference reflector 4, placed at a fixed distance in front of the lightguide output. Images of the reflector, sequentially illuminated by the selected single wavelengths, are taken by a digital camera 5 placed at a fixed distance from the reflector.

To conclude, the snapshot multi-spectral-line imaging (SMSLI) method described in this chapter appears relatively simple and potentially easy to implement. Thanks to the narrowing of both spectral and temporal intervals of imaging, this method could notably improve the performance of MSI. The common RGB color cameras, including those embedded in smartphones, are well-suited for snapshot capturing of three ultra-narrowband spectral images; four to nine such images could be captured within sub-second time intervals by switching the sets of illumination wavelengths [15]. The mosaic multiband image sensors enable capturing an even larger number of spectral line images (e.g., 16, 32) by a single snapshot under appropriate spectral line illumination. Generally, uniform multi-spectral-line illumination is a key factor for successful ultra-narrowband multispectral imaging; some design solutions for this kind of illumination using lasers will be discussed in the next chapter.

CHAPTER 3

Multilaser Illumination Designs

I F LASERS ARE SELECTED as the spectral line emitters, one of the main challenges for the implementation of the SMSLI method is ensuring uniformity of the whole target area illumination by each of the applied laser wavelengths. If not fully uniform, at least the same distribution of illumination intensity at all exploited wavelengths is required. The commercial RGB laser projectors intended for laser color TV, laser shows, and other artistic applications are generally suitable as illuminators for snapshot triple-wavelength imaging in the visible range. However, such devices are expensive and bulky, designed mainly for stationary settings. If aiming at mobile applications of SMSLI, more compact and flexible multilaser illumination designs are preferable. Several design options developed over the recent decade will be presented and discussed in this chapter.

3.1 JOINT OPTICAL FIBER ILLUMINATOR

In paper [9], dual-wavelength spectral line imaging was performed by means of the RGB camera under simultaneous illumination by two milliwatt-range laser spectral lines with wavelengths

DOI: 10.1201/9781003476702-4

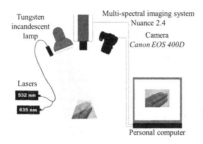

FIGURE 3.1 Setup scheme for dual laser spectral line illumination by means of Y-shaped optical fiber configuration (left), in parallel with white light illumination for hyperspectral imaging (middle) [9].

532 and 650 nm. To ensure similar distributions of illumination intensities at both wavelengths, a simple optical setup comprising Y-shaped double optical fiber assembly was developed. Each laser beam was coupled into an optical fiber with silica core diameter 400 μm and numerical aperture NA = 0.22 (Figure 3.1). The distal ends of both fibers were flat polished and tightly joined together in a cylindrical ferrule, so both output beams had practically the same optical axis. The fiber's output radiation was further expanded by a concave lens to illuminate the target area of ~5 cm diameter. Although the uniformity of target illumination was not ideal, the illumination intensity distributions at both wavelengths, being mainly determined by the optical fiber's numerical aperture, were nearly the same. The measurements using white paper as a target confirmed that the pixel's output signal ratios at both wavelengths over the whole illuminated area were practically equal, which ensured a successful dual spectral line imaging procedure as described in Paragraph 2.1. Generally, this dual laser-fiber design concept may be further extended for three, four, or more spectral line illumination, using the corresponding number of joint laser-coupled optical fibers, each emitting at a different wavelength.

3.2 RING-SHAPED OPTICAL MULTIFIBER ILLUMINATOR

The above-regarded design suggests coupling of each illumination laser beam into a single optical fiber. Another option of multilaser forward illumination – coupling the laser beams into multifiber bundles with ring-shaped arrangement of the fiber distal ends – was proposed and tested in [10,11,16]. The illumination design scheme is presented in Figure 3.2. Emission of three stabilized cw laser modules emitting blue, green, and red spectral lines with output powers in the range of 10–15 mW was launched into the respective three SMA-type connectors, each comprising a bundle of seven fibers with silica core diameter 400 μm. Their output ends (altogether 21) were randomly distributed and fixed in a polished metallic ring surrounding the CMOS camera, used for the triple spectral line imaging. With all lasers switched on, the ring emitted simultaneously three spectral lines. Passing through a diffusing film, the fiber's ring output light provided even triple-wavelength illumination of a round target area with a diameter of 30 mm, located 5 cm apart from the camera objective lens. The uniformity of target illumination was better if compared to the single-fiber coupling design described in the previous paragraph. Again, this illumination design can be

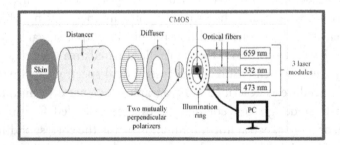

FIGURE 3.2 The design scheme for triple laser wavelength illumination by a ring-shaped optical fiber bundle assembly [10,17].

further extended for four, five, or more spectral line illumination just by increasing the number of emitting lasers and coupled to them optical fiber bundles.

3.3 BACK-SCATTERED LASER BEAM ILLUMINATOR

A compact, smartphone-compatible three wavelength illuminator design based on back-scattering of the laser radiation has been proposed and described in [18,19]. The illumination unit comprises six laser modules and a ring-shaped specially designed collector-diffuser for laser beam management. Figure 3.3 shows the general design scheme (left) and the optical scheme for the collector-diffuser (right). The illumination spectral lines at wavelengths 448, 532, and 659 nm are emitted by three couples of compact 20 mW power laser modules. Every pair of lasers emits precisely the same wavelength, guaranteed by the manufacturer and tested in lab. Laser modules of each equal-wavelength pair are mounted at opposite sides on the internal wall of a hollow shielding cylinder; the round bottom opening of this cylinder (diameter 40 mm) is in contact with the target (e.g., skin) and forms the field of view for the smartphone camera "looking" through the central opening of the upper disk.

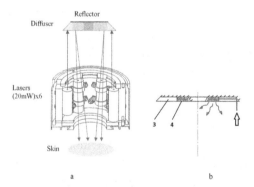

FIGURE 3.3   The design scheme of the back-scattering triple laser wavelength illuminator (a) and light propagation in the laser beam collector-diffuser (b) [18,19].

All six coaxial laser beams are pointed to the 45° conical reflecting edge of a flat transparent ring 3 (Figure 3.3b) acting as a beam collector. After reflections, all laser beams are turned radial toward the internal ring-shaped diffuser 4. The arrow in Figure 3.3b denotes one of six co-axial laser beams that are further scattered by the internal ring 4 made of milky plexiglass. The upper and side surfaces of the laser beam collector/diffuser unit are mirrored. In result, the flat diffuser 4 evenly illuminates the target area simultaneously by all three laser wavelengths.

Alternatively, the external ring 3 can be replaced by a set of radially oriented flexible optical fibers or other appropriate light guide(s) that deliver the laser radiation to the diffusive ring 4 from the laser modules that are placed elsewhere.

If compared with the laser illumination methods that exploit beam-expanding or scattering elements placed in between the laser source and the target area, the above-described design provides more uniform illumination of the selected tissue surface because the diffuser 4 acts as an isotropic surface emitter, not as a point-source.

The uniformity of target illumination was checked experimentally at all combinations of illumination wavelengths (448, 532, 659 nm) and image sensor detection channels (R, G, B), using white paper as the target. Figure 3.4 illustrates the obtained illumination intensity distributions, detected at separated RGB output channels in two cases – (i) when only one wavelength was used for illumination (659 nm detected at R-channel, 532 nm detected at G-channel and 448 nm detected at B-channel, the upper row), and (ii) when all three wavelengths of illumination were simultaneously switched on (the lower row). One can conclude that this design provides good multilaser illumination uniformity and is well suited for triple spectral line imaging. One can also mention that the design is not limited to triple spectral line illumination but allows extension of the number of illumination spectral lines to four, five, or more just by the increased number of laser couples emitting the same working wavelength.

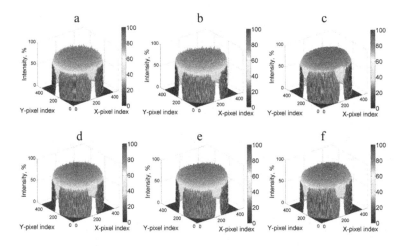

FIGURE 3.4 Uniformity of white paper illumination at different illumination-detection combinations. (a) R-image at single-wavelength 659 nm illumination; (b) G-image at single-wavelength 532 nm illumination; (c) B-image at single-wavelength 448 nm illumination; (d–f) R-, G-, and B-images at simultaneous 3-wavelength illumination. Scale on the right represents relative intensities. Diameter of the illuminated area – 40 mm [19].

3.4 SIDE-EMITTING OPTICAL FIBER ILLUMINATOR

Side-emitting optical fibers, including those with silica core [20,21], are flexible linear light sources being increasingly used for light framing and various artistic applications [22]. The unique properties of such fibers make them very attractive also for illumination simultaneously by several laser spectral lines [23]. Besides flat targets, side-emitting fiber formations can evenly illuminate curved surfaces, as well.

Figure 3.5a illustrates one of the options where the radiation emitted from laser 1 (which may include several discrete spectral lines, for example, with wavelengths λ_1, λ_2, and λ_3) is launched via connector 2 into the side-emitting optical fiber 3. The fiber can be bent into one or more loops to evenly illuminate the object's

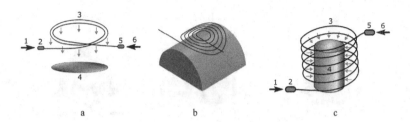

FIGURE 3.5 Geometries of multilaser surface illumination by side-emitting optical fiber loops: (a) flat target, (b) curved surface of the target, (c) cylindrical surface of the target [23].

surface 4. Through connector 5 installed at the other end of the fiber, radiation from another laser 6 emitting a different set of discrete wavelengths (for example, λ_4, λ_5, and λ_6) can be launched into the fiber. In result, six laser lines simultaneously illuminate the target area.

The folded part of fiber 3 can form either a plane configuration (circle, ellipse, spiral, texture, etc.) or a three-dimensional figure. In the latter case, the shape of illuminating fiber formation can be matched to the shape of a curved surface so ensuring even and shadowless illumination by all laser wavelengths (Figure 3.5b and c).

The proposed solution has several advantages compared to the other known designs. First, the size of illuminator is significantly reduced thanks to the small dimensions of side-emitting optical fibers (typical external diameter ~1 mm, minimum loop diameter ~20 mm), which ensures compactness of the device. Second, the ring-shaped (or other shaped) side-emitting fiber is a continuous emitter, i.e., all its points emit simultaneously, which provides better quality of illumination than in the case of discrete light sources – for example, the ring-shaped output ends of conventional optical fibers or laser diode rings. In such cases, each quasi-point light source individually affects the quality of the object's surface illumination. Third, the distribution of surface illumination intensity, even if sometimes not being perfectly uniform, is always identical for all used laser spectral lines, as the illuminator's geometry is the same for all spectral lines launched into the side-emitting optical fiber. Fourth, the optical fiber's flexibility makes it possible to easily change the configuration of the illuminator both in the plane and in space. The same side-emitting fiber setup can be modified for uniform multiline illumination of both flat and curved surfaces. Fifth, the proposed solution can provide uniform illumination from all sides of spatial object surfaces (4π-illumination) – see an example for a cylindrical surface in Figure 3.5c. Finally, side-emitting fiber light source is electrically safe and can be placed in liquids or harsh environments.

3.5 REMOVAL OF LASER SPECKLE ARTIFACTS IN SPECTRAL LINE IMAGES

The term *speckle* refers to a random granular pattern which can be observed when a coherent laser beam is diffusely reflected at a non-mirror surface with a rough structure, such as paper or white wall paint. This phenomenon results from the interference of many different reflected portions of the incident beam with random relative optical phases. If lasers are applied as illumination sources for SMSLI, the recorded spectral line images usually contain the speckle artifacts or "grains" in the images. Still, this sort of image defect can be minimized or fully avoided.

One of the possible solutions is proposed in [24]. To prevent the appearance of speckle artifacts in spectral images, mechanical vibrations of the light-diffusing element were found to be efficient. The proposed device (Figure 3.6a) comprises a digital image sensor 2 having an objective lens adapted for object 1 imaging, a laser irradiation unit 3, a reflector 4 with a light scattering coating 5, connected to a mechanically vibrating element 6, and the oscillation generator 7. Several accordingly aligned lasers and their powering devices are mounted in this unit 3. Generator 7 provides an adjustable oscillation frequency f which satisfies the condition $t.f > 5$ (where t – exposure time of the image sensor 2), so providing at least five oscillations of the scattering reflector during a single exposure. Considering the sub-micron wavelength of the laser

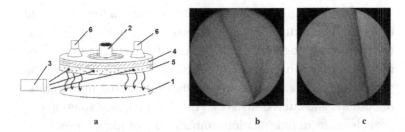

FIGURE 3.6   Design scheme of a device preventing laser speckle artifacts in the spectral line images (a) and a spectral line image before (b)
and after (c) switching on the vibrations with frequency 80 Hz; exposure
time 0.1s [24].

radiation, even the oscillation amplitude of one micron essentially
changes the diffused laser radiation interference conditions on
the object's surface. As a result, the laser speckles are periodically
moving along the surface. This leads to speckle "smearing" during
the image capturing exposure, so reducing or eliminating graini-
ness of the image and thereby improving its quality.

The dimensions of vibrating components can be very small – for
instance, when using a mm-sized smartphone speaker with
attached lightweight metallic diffusing foil to its moving dia-
phragm. The effectiveness of such design is illustrated in Figure 3.6b
where RGB images of skin captured by the smartphone camera at
green laser 532 nm illumination are compared with steady (b) and
oscillating (c) reflector-diffuser. The image quality considerably
improves when the proper vibrations take place.

Only one of many possible designs for removing laser speckle
artifacts in the spectral images was discussed here. Generally,
any moving or vibrating item on the laser illumination pathway
(e.g., a vibrating optical fiber that transmits the laser radiation)
minimizes or fully removes the speckle artifacts, if the camera
exposure time is sufficiently long.

To conclude, several possibilities of uniform illumination
of flat or curved target surfaces simultaneously by several laser

spectral lines have been discussed in this chapter. Four appropriate illumination designs were described, along with a method for the reduction/removal of the laser speckle artifacts in the spectral line images. These or similar designs in conjunction with digital image sensors can be exploited for the development of high-performance multispectral imaging devices able to obtain sets of ultra-narrowband spectral images by the methodology described in Chapter 2. If all four designs are mutually compared, the side-emitting fiber illuminators appear to be the most promising for future developments thanks to the flexibility, small dimensions, and high uniformity of flat or curved target illumination simultaneously by several laser lines.

Prototype Devices for Multi-Spectral-Line Imaging

A NY IDEA IS VIABLE only when practically implemented. Even if the main concepts and design ideas seem to be fully convincing, they must be experimentally validated. In the case of SMSLI technology, it means that functioning ultra-narrowband multi-spectral imaging devices should be designed, assembled, and tested. Several custom-made experimental "proof-of concept" prototypes for multispectral-line imaging implementing the above-presented illumination designs have been assembled and tested at the author's lab; ten of them will be described in this chapter.

4.1 TRIPLE-WAVELENGTH 3 × 7 OPTICAL FIBER PROTOTYPE

In Paragraph 3.2, a triple laser illumination design comprising three bundles of seven optical fibers each with their 21 outputs arranged in the shape of a ring was presented. Our first experimental prototype device implementing this design [10] is shown

DOI: 10.1201/9781003476702-5

FIGURE 4.1   Outlook of the triple laser – 21 optical fiber ring illumination SMSLI prototype device [10].

in Figure 4.1. It consists of a CMOS camera (USB 2 UI-1226LE-C from IDS Imaging Development Systems GmbH, max. 87 frames/sec in VGA video-mode) with an objective (Lensagon BM8018S118ND from Lensation GmbH), surrounded by a triple laser wavelength illumination ring. Emission of three stabilized cw laser modules (405 nm, DE405-12-5; 532 nm, DD532-10-5; 650 nm, DB650-12-3, all from Huanic Corp., CN) with output powers in the range 10…15 mW is launched into the corresponding three SMA-type connectors, each comprising a bundle of seven fibers with silica core diameter 400 μm. Their output ends are randomly distributed around the CMOS objective, forming the ring composed of 21 emitting fiber ends for illumination of a round, 30 mm in diameter flat target area, located 5 cm from the CMOS objective lens. To ensure better uniformity of illumination, a conical white plastic diffuser, serving also as a spacer, is placed between the ring and the target.

During the measurements, all automatic features of the RGB sensor are switched off, and the output powers of all lasers and the exposition time of CMOS sensor are adjusted to provide linearity of photovoltaic response at all spectral combinations. The three laser modules are powered by standard laboratory power supplies, not shown in the figure. The camera's electronic board (in the upper part of Figure 4.1) is also connected to a standard power supply.

In some later experiments, the 405 nm laser module has been replaced by another one emitting 473 nm, and the 650 nm module – by a module emitting the 659 nm spectral line [16].

4.2 SMARTPHONE-COMPATIBLE TRIPLE LASER LINE PROTOTYPE

The above-described prototype with external power supplies functioned well in laboratory experiments, but it is not convenient for mobile applications. Therefore, a compact battery-powered and smartphone-compatible prototype device implementing the illumination design described in Paragraph 3.3 has been developed. Figure 4.2 shows its design details (left) and outlook of an operating prototype with a smartphone on it (right). The illumination wavelengths 448, 532, and 659 nm are emitted by three pairs of compact 20 mW power laser modules (models PGL-DF-450nm-20mW-15011564, PGL-VI-1-532nm-20mW-15030443, and PGL-DF-655nm-20mW-150302232 from Changchun New Industries Optoelectronics Tech.

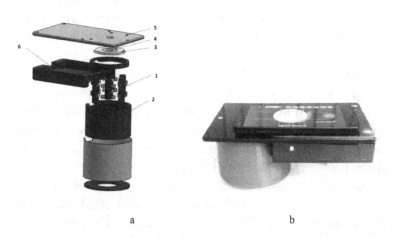

a   b

FIGURE 4.2   3D model of the 3-wavelength/6-laser illumination unit (a) and outlook of the mobile prototype with smartphone on it (b). 1 – laser modules (3 pairs, 448-532-659 nm), 2 – shielding cylinder, 3 – collector of laser beams, 4 – flat ring-shaped diffuser of laser light, 5 – sticky platform for the smartphone, 6 – electronics compartment [19].

Co., Ltd). The six laser modules 1 are mounted on the internal wall of a hollow 3D-printed plastic shielding cylinder 2; the round bottom opening of this cylinder (diameter 40 mm) is in contact with the target and forms the field of view for the smartphone camera, situated 80 mm apart. The transparent disc 3 (beam collector) with the 45° conical reflecting edge is made of a standard Plexiglas, while the internal ring-shaped flat diffuser 4 is made of milky Plexiglas. The dimensions of the collector-diffuser unit 3–4 are as follows: external diameter 50 mm, inner-outer diameters of the diffusing ring 10–40 mm, thickness 3 mm; the upper and side surfaces of this component are mirrored by a vacuum-sputtered Al-coating.

The smartphone – model Google Nexus5 comprising 8 Mpx image sensor SONY IMX179 with known (provided by the manufacturer) RGB-sensitivities – is placed on a flat sticky platform 5 with a round window for the smartphone rear camera, co-aligned with the internal opening of diffuser 4. The sticky platform design allows using nearly any model of smartphones or tablet computers, independently of their dimensions. This prototype has been mainly used for skin diffuse reflectance imaging; to avoid detection of skin surface-reflected light by the smartphone camera, cross-polarizer system is used: the round camera window is covered by a film polarizer, and another film with orthogonal direction of polarization covers the bottom of diffuser 4.

The microcontroller STM32F103 manages operations of the three laser power modules via control of the Laser Power Supply Unit (PSU). Each pair of laser modules can be switched on and off independently by means of the control buttons. Two rechargeable batteries (Samsung 18650 3.7 V Li-Ion, 2.8 Ah) and laser power supply circuits are placed in the 3D-printed plastic compartment 6 below platform 5. Target illumination intensity is adjusted to ensure linearity of photo-detection by the smartphone camera. Its automatic settings are switched off by the *AZ Camera* software, so the R-, G-, and B-outputs of the image sensor responded accordingly to the spectral sensitivities of its three detection bands.

4.3 DUAL-CAMERA FOUR SPECTRAL LINE SNAPSHOT IMAGER

Two prototype devices for snapshot capturing of four spectral line images have been developed – one with two cameras aiming at the same target and the other with a single four-band camera. Here the first of them [25] will be discussed. The design scheme of this device is presented in Figure 4.3. Two slightly tilted cameras (RGB and NIR, MQ022CG-CM and MQ022RG-CM from Ximea, DE) equipped with 425 and 800 nm long-pass filters (models #84-742 and #66-235 from Edmund Optics, UK) are simultaneously capturing images of the same round target area of diameter 30 mm, with subsequent extraction of four spectral line images. The RGB camera detects three visible spectral line images (at 445, 523, and 628 nm) while the NIR camera detects the 850 nm spectral line image.

Two laser modules – RGB fiber-coupled module (Elite Optoelectronics, CN) emitting 20 mW at each of the three

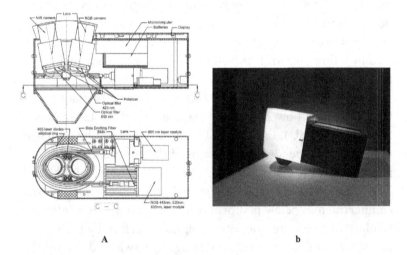

A   b

FIGURE 4.3   Design scheme (a) and outlook (b) of the dual-camera/four wavelengths prototype [25].

visible spectral lines and a tunable power 850 nm/40 mW module (RLDH850-40-3 from Roithner, AT) – are used for four wavelengths illumination. Several semi-elliptical loops of side-emitting 400-μm silica core optical fiber (from Light Guide Optics Int., LV) are exploited as the light source for even multilaser illumination of the target area. The side-emitting fiber is SMA-terminated at both ends; one of them is used for the RGB laser input and the other one for the 850 nm laser input. In result, the examined target (e.g., skin) area is uniformly illuminated by the four above-mentioned laser spectral lines. Besides, four 405 nm laser diodes for excitation of the target fluorescence are squarely placed in the middle zone of the illuminator.

The device is initiated by the operator pushing the START button and selecting the appropriate operation regime (SMSLI only, fluorescence image only, combined mode) and exposure times for the RGB and IR cameras. Then both cameras and RGB lasers are switched on and the device is properly placed on the target area (monitored on the flat display). After pressing the SHOT button, micro-controller (STM32G071 from STMicroelectronics, CH) synchronizes proper illumination of the target to capture the image (or set of images) that is/are read by the single-board-computer (SBC, Rock960 from Vamrs, CN). The recorded image is transferred to the display (5.5inch HDMI AMOLED from Waveshare, CN) and/or transmitted via *Wi-Fi* to the remote computer for image processing. The device is fully self-sustained by using rechargeable Li-ion batteries (INR18650-25R from Samsung, KR) as the power supply.

4.4 SINGLE CAMERA FOUR SPECTRAL LINE SNAPSHOT IMAGER

The above-described device functioned well, but only if the target area was ideally flat. In the case of skin imaging, some anatomy-determined curvatures of the target zone are possible, and then the images taken by both cameras can differ. To avoid this disadvantage, a prototype device of a very similar illumination design but comprising only one camera with four sensitivity bands was developed [14].

The design scheme of the device is presented in Figure 4.4, and its functional block-diagram is presented in Figure 4.5. The optical setup is nearly the same as that of the double-camera device: it includes the side-emitting fiber loop for illumination, RGB and NIR lasers, and the other components. In this device, they are packed more densely (therefore a cooling fan is added) and the overall design is handier for mobile applications. The target area is round with a diameter of 10, 20, or 30 mm, depending on the used changeable conical nozzle; each nozzle is inside-covered by a black coating film (Spectral Black from Actar, IL). The electronic solutions are somewhat different from those of the double-camera device; four rechargeable Li-ion batteries (INR18650-35E from Samsung, KR) supply power, so the device is fully self-sustained.

Four-band RGB-NIR camera (MSC-RGBN-1-A from Spectral Devices Inc., CA) captures images of the targeted area with subsequent extraction of the four spectral line images at 445, 520, 635, and 850 nm (see Figure 2.7). The image sensor is covered by a 420 nm long-pass filter (to cut off the 405 nm laser radiation used for fluorescence excitation) and equipped with an objective lens

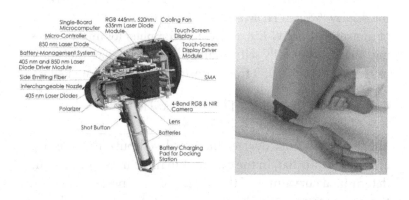

a   b

FIGURE 4.4   Design scheme of the single camera prototype for snapshot four spectral line imaging (a) and its outlook (b) [14].

(25 mm #67–715 from Edmund Optics, US) with a top-mounted VIS-NIR polarizer.

The four spectral line images are taken by a single snapshot, and they are stored in the embedded single-board computer (SBC, Rock960 from Vamrs, CN) and Wi-Fi transmitted to an external computer for further calculations. Within a second, the laser modules are switched off, violet 405 nm diode lasers are switched on, and a fluorescence image at the G-channel of camera is captured. This image is transmitted to the external computer, as well. This last step can be skipped or used alone.

The device is initiated by the operator pushing the START button and selecting the appropriate operation mode (SMSLI only, fluorescence image only, or the combined mode) and exposure time for the RGB-NIR camera. Then the camera and lasers are switched on, and the device is properly placed on the skin target area (monitored on the round display). After pressing the SHOT button, the micro-controller (STM32G071 from STMicroelectronics, CH) synchronizes proper illumination of the target to capture the image (or set of images) which is/are read by the single-board-computer. The recorded images are seen on the round display and can be transmitted via *Wi-Fi* to the remote computer for additional calculations, e.g., calculations of the target pigment distribution maps. The recorded RAW images (2048×2048 pixels) are stored in the computer hard disk.

4.5 PROTOTYPES FOR SEQUENTIAL SPECTRAL LINE IMAGING

The single-snapshot RGB technique is not applicable for obtaining more than three spectral line images. Still, a sequential double- or triple-snapshot approach can be applied to capture a higher number of ultra-narrowband spectral images (Paragraph 2.3). This concept was first implemented in a prototype device comprising a switchable four laser illuminator and a smartphone (Figure 4.5). Two of the laser modules could be manually re-switched, so providing two sets of three wavelengths illumination: 405/532/650 nm and 450/532/650 nm. Four rechargeable AA-type batteries were

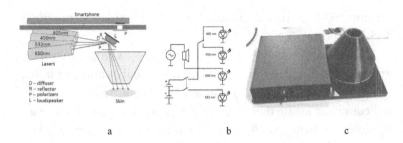

FIGURE 4.5   Design scheme (a), electronic circuit (b), and outlook (c) of the prototype for manually switchable four wavelengths laser illumination [26,27].

used for power supply. Relatively uniform illumination of a round spot (diameter 18 mm) was provided by an advanced optical design which also reduced laser speckle artifacts by means of loudspeaker-initiated vibrations [24,27].

As the next step, a portable self-sustained automatically switchable five wavelength (405, 448, 532, 659, and 842 nm) laser module illuminator of similar optical design, additionally equipped with own NIR-sensitive CMOS camera, Raspberry Pi3 processing unit and 3.5′ touch screen, has been designed and assembled (Figure 4.6). Two consequent RGB snapshots are to be taken for spectral image acquisition: the first when 405, 532, and 659 nm lasers are switched on and the second when 448, 532, and 842 nm lasers are switched on.

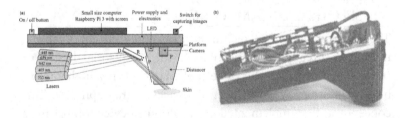

FIGURE 4.6   Design scheme (a) and outlook (b) of the prototype device with automatically switchable 5-wavelengths illumination: R – reflector, D – diffuser, P – polarizers [26].

 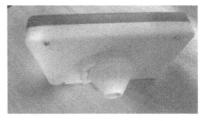

FIGURE 4.7   Front and rear views of the second prototype for obtaining five spectral line images [26].

This design concept was further developed, introducing a five-wavelength laser diode ring as a light source (which ensured better uniformity of skin illumination at all wavelengths) and white LED illumination for capturing color images. This prototype version also comprised a larger (5 inch) display as shown in Figure 4.7. The cornerstone of this device was the Raspberry Pi minicomputer. It captured the images taken with a camera module comprising NIR-sensitive OmniVision OV5647 image sensor and saved the images to a microSD card that also comprised the operating system of the device. Five different wavelength laser diodes (405, 450, 520, 660, 850 nm) were used for illumination; extra white LED source ensured correct pointing to the region of interest under a display control. To reduce laser speckle detection, a vibrating speaker was attached to the diffuser. Built-in drivers ensured the proper functioning of all devices.

Based on the gained experience, one more design concept was implemented with a ring-shaped illumination source comprising five sets of laser diodes [28,29]. The illumination part involves four groups of five laser diodes (20 laser diodes in total) emitting at the wavelengths 405 nm (20 mW), 450 nm (80 mW), 525 nm (50 mW), 655 nm (15 mW), and 845 nm (50 mW), all from Roithner Lasertechnik GmbH. Four white LEDs (1W) illuminate the skin in preview mode. The diodes are soldered on 1 mm thick PCB ring that is surrounding the camera lens and fixed in the front part of the housing – see Figure 4.8.

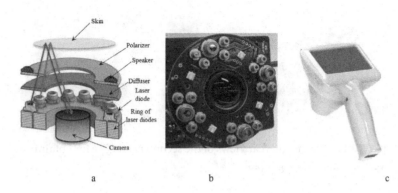

FIGURE 4.8 The optical design scheme of the 5×4 laser diode device (a), its technical implementation (b), and outlook of the prototype (c) [28,29].

1/2-Inch 3-Megapixel CMOS Digital Image Sensor MT9T031 from Aptina with pixel size 3.2 μm and a built-in 10-bit ADC providing 1,024 grades of intensities is used for image capturing via linear polarizing film, oriented orthogonally to that in front of the illumination ring. The main processing unit is 1 GHz dual core Tegra 2 module from Toradex AG, running the Windows CE7 operating system. STM32L4 series Cortex M4 microcontroller is connected to low-speed peripherals and performs its tasks in real time. User interface, measurement process, and results are displayed on a 4.3″ MVA-type capacitive touch screen from Newhaven Display International, Inc.; lithium-ion accumulator is placed in the device's grip. The power unit of the device consists of a voltage boost and current regulation circuit for lasers, an integrated battery management circuit (regulation and charging), a voltage stepdown converter for the entire system (CPU, MPU), and an ultra-low noise, linear, low-dropout voltage regulator for the CMOS image sensor [29].

4.6 PROTOTYPE FOR LARGE AREA SKIN TRIPLE SPECTRAL LINE IMAGING

Whole-body imaging systems recently have gained attention in dermatology as efficient tools for detection, diagnostics, and monitoring of various skin malformations, including cancers.

Contrary to classic investigations focused on one lesion, when a single image is taken using a dermoscope or photo-camera, such systems capture several high-resolution photographs of the whole body or its specific areas. This allows not only detection of lesions but also comparing the images acquired at different times and identifying the expanding malformations – potential melanomas. The existing commercial systems use broadband white light sources for capturing color images of the skin; spectral features of the malformations remain hidden. To ensure more advanced diagnostics, a relatively simple and less expensive prototype for spectral imaging of large area *in vivo* skin under triple laser line illumination has been developed [30,31].

The light source of the prototype (Figure 4.9a) is an elliptic spiral made of a 60 m long silica core side-emitting fiber (GF-600 from Light Guide Optics Int., LV), mounted on a flat mirrored surface with rectangular inside opening for 61 Mpx RGB camera (Sony a7R IVA with the Sony SEL2470GM, F2.8 G Full Frame Standard Zoom Lens). The fiber at one end is SMA-terminated and connected to the RGB laser (3IN1 RGB High Power 3 W White Laser from NaKu Technology Co. Ltd, CN) which emits simultaneously three spectral

a   b

FIGURE 4.9   The prototype for triple spectral line imaging of large area or whole-body skin (a) and its placement in a light-shielding tent (b) [31].

lines at the wavelengths 450, 520, and 638 nm, at up to 1 W power each. A reflector mounted at the other end of fiber ensures relatively uniform side emission over the entire fiber length [21]. Both the fiber illuminator and the camera are mounted on a platform that can be automatically moved to the upper (~1.5 m) and lower (~0.5 m) positions for capturing three spectral line images of the upper and lower parts of the patient's body at various positions. To avoid illumination by ambient light, the prototype and patient are shielded in a light tent, serving also as the patient's dressing cabin (Figure 4.9b). The device is operated from outside of the tent via Wi-Fi by an external computer with dedicated software ensuring both image capturing and vertical displacement of the camera-illuminator system. All obtained large-area, high-resolution skin spectral line images are stored in the computer memory for further processing, including identification, numbering, and sorting of the detected skin lesions, creation of album with lesion's images, and calculations of the distribution maps for content increase/decrease of three main skin chromophores.

4.7 PROTOTYPE FOR ENDOSCOPIC TRIPLE SPECTRAL LINE IMAGING

Filtering relatively narrow (tens of nanometers) spectral bands out of the white light endoscopic illumination facilitates contrast of specific tissue features in the endoscopy images; the NBI (narrow band imaging) technique has been widely used by endoscopists over the recent years. Implementation of the SMSLI technology in endoscopy might improve the spectral selectivity (color sensitivity) by narrowing the imaged spectral bands down to the laser linewidth. Besides, selective triple wavelength illumination enables spectral evaluation of the detected mucosal pathologies.

To explore this challenge, a prototype setup shown in Figure 4.10 has been assembled and tested [30,31]. It comprised RGB laser emitting the 450, 520, and 638 nm spectral lines of manually adjustable power (300 mW White Laser from NaKu Technology Co. Ltd, CN) and an intra-nasal endoscope

FIGURE 4.10 The prototype setup for endoscopic triple spectral line illumination by RGB laser [31].

(STORZ rigid model HOPKIN II or Rhino-Laryngo-Fiberscope, model 11001RDK1) with the RGB camera TH110. A specially designed flexible metal-armored optical fiber cable with a vibrating element was used for laser light delivery; it was of the same length as the original lightguide used with this endoscope, so equal comfort of operation was ensured. The RGB laser light was launched through the input of built-in illumination channel of the endoscope via an adapted optical fiber connector, while the laser radiation was launched via the standard SMA connector mounted at the other end of the fiber cable. This illumination design did not require any additional sterilization measures and ensured a similar illumination geometry as the original white light LED illumination system. By appropriate adjusting of the emission power for each of the three laser lines, their integral white illumination at the distal end of endoscope was set to have the same or smaller intensity as that using the original broadband white light source, so also from this point conditions for endoscopists remained unchanged.

To conclude, the viability of the snapshot multi-spectral-line imaging (SMSLI) concept was demonstrated in this chapter on examples of ten functioning lab-developed prototype devices. They have passed or undergo validation tests in laboratory and/or clinical environments to collect unique sets of ultra-narrowband spectral images and, in parallel, to find out specific issues to be improved for reaching the next level, i.e., commercial prototypes. Most of the regarded devices are primarily intended for clinical applications which demand extensive clinical trials with statistically significant number of patients and/or volunteers. It is challenging as such trials are possible only if external investors and commercial partners are attracted; university labs have not enough resources for this kind of activity.

CHAPTER 5

Applications for Skin Diagnostics

The SMSLI method and proof-of-concept experimental prototypes described in the previous chapters were clinically validated,
mostly focusing on advanced diagnostics of skin malformations,
including skin cancers. Sets of spectral line images can be con-
verted into distribution maps characterizing the content changes
of the main skin chromophores (melanin, oxy-hemoglobin, and
deoxy-hemoglobin) in skin lesions. Such maps help dermatologists with proper diagnoses and follow-up of the healing pro-
cesses. Several hundred clinical cases have been examined with
the above-described devices, starting with the proof-of-concept
studies [17,32] using our first laboratory prototype described in
Paragraph 4.1 and ending up with the whole-body multispectral
imaging system presented in Paragraph 4.6.

5.1 METHODOLOGY FOR CONVERSION OF SPECTRAL LINE IMAGES INTO SKIN CHROMOPHORE MAPS

The basic approaches for mapping of the three main skin chromophore variations in malformations by means of the SMSLI technique are described in [19]. The general concept is illustrated in Figure 5.1. Let us suppose that RGB color image of a skin pathology spot, surrounded by healthy skin, is captured under illumination that comprises only three equal intensity spectral lines at wavelengths λ_1, λ_2, and λ_3 (the vertical lines in Figure 5.1). With respect to the spectral sensitivity of the RGB image sensor and the crosstalk between its detection bands at the working wavelengths, three monochromatic spectral images are extracted from the color image dataset by the technique described in Chapter 2. The three-chromophore skin model suggests that the dominant absorbers in the skin at any of the fixed wavelengths λ_j ($j=1, 2, 3$) are oxy-hemoglobin (further abbreviated by a), deoxy-hemoglobin (b), and melanin (c) – see the crossings of their absorption curves with the vertical lines in Figure 5.1. If the skin surface reflection is suppressed by means of crossed polarizers, variations in chromophore composition would lead to changes in the diffusely reflected light intensities at each of the fixed wavelengths. Such variations in the pathology region relative to the healthy skin can

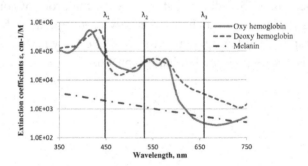

FIGURE 5.1 Absorption of three main skin chromophores at three fixed wavelengths 448, 532, and 659 nm [19].

be estimated by measuring reflected light intensities from equally sized regions of interest in the pathology (I_j) and the adjacent healthy skin (I_{jo}). The calculated ratios I_j/I_{jo} at each pixel or pixel's group of three monochromatic spectral images contain information on the concentration increase or decrease of all three regarded chromophores Δc_i ($i = a$, b, c), which can be further mapped over the whole image area. Our experience confirms that healthy skin around the pathology is a more reliable and handier reference than a calibrated white reflector.

The above-mentioned intensity changes at three exploited wavelengths due to chromophore's absorption can be represented in terms of the Beer–Lambert–Bouguer law [33,34]:

$$
\left\{
\begin{aligned}
\ln\left(\frac{I_1}{I_{01}}\right) &= -l_1\left(\Delta c_a \cdot \varepsilon_a(\lambda_1) + \Delta c_b \cdot \varepsilon_b(\lambda_1) + \Delta c_c \cdot \varepsilon_c(\lambda_1)\right) \\
\ln\left(\frac{I_2}{I_{02}}\right) &= -l_2\left(\Delta c_a \cdot \varepsilon_a(\lambda_2) + \Delta c_b \cdot \varepsilon_b(\lambda_2) + \Delta c_c \cdot \varepsilon_c(\lambda_2)\right) \quad , \\
\ln\left(\frac{I_3}{I_{03}}\right) &= -l_3\left(\Delta c_a \cdot \varepsilon_a(\lambda_3) + \Delta c_b \cdot \varepsilon_b(\lambda_3) + \Delta c_c \cdot \varepsilon_c(\lambda_3)\right)
\end{aligned}
\right.
$$

$$(5.1)$$

where $\varepsilon_i(\lambda_j)$ are extinction coefficients of three regarded chromophores at three exploited wavelengths, and l_j is the mean path length of remitted photons in skin at a particular wavelength. The increase or decrease in chromophore concentration at each image pixel (or selected group of pixels) is found by solving the linear equation system (equation 5.1) with abbreviated measured quantities $k_j = \ln(I_j/I_{jo})$:

$$
\left\{
\begin{aligned}
\Delta c_a &= A_1 \cdot k_1 + A_2 \cdot k_2 + A_3 \cdot k_3 \\
\Delta c_b &= B_1 \cdot k_1 + B_2 \cdot k_2 + B_3 \cdot k_3 \, . \\
\Delta c_c &= C_1 \cdot k_1 + C_2 \cdot k_2 + C_3 \cdot k_3
\end{aligned}
\right.
\qquad (5.2)
$$

The coefficients A_j, B_j, and C_j comprise numerical values of the corresponding chromophore extinction coefficients $\varepsilon_i(\lambda_j)$ [7,35] and the remitted photon path lengths l_j at the three exploited wavelengths.

The Beer–Lambert–Bouguer law originally relates to the absorption of a light beam passing through a slab of substance with thickness x. If this law is adapted to the diffusely remitted light geometry, slab thickness x must be replaced by the mean path length l_j of the skin back-scattered photons at a particular wavelength. To estimate the l_j values with respect to light scattering properties, usually Monte Carlo simulations in the frame of some selected skin models are performed [19]. However, real skin structure at specific body locations may not correspond to the model assumptions and therefore could lead to mistaken results. Besides, the common models consider all launched photons (including those absorbed and forward-scattered), while only the back-scattered photons should be regarded as they are detected by the optical contact probes and imaging cameras of the clinical devices. Clearly, directly measured remitted photon path length distributions and mean values are preferable for practical use.

Such measurements were performed in our study [36]. In this work, picosecond laser pulses at seven different wavelength bands were launched into healthy human forearm skin via a specially designed optical fiber contact probe that enabled measurements of the skin-remitted pulses at five different distances from the input fiber. The shapes of input and output pulses were compared to determine the mean time-of-flight values of the skin-remitted photons and the related mean photon path lengths at all 35 spectral-spatial combinations.

The measurement setup (Figure 5.2) comprised a broadband picosecond laser (Fianium from NKT PHOTONICS, DK) as the pulsed light source and a time-correlated photon counting system – photomultiplier HPM-100-07 combined with the controller DCC-100 and data processing card SPC-150 (all from Becker&Hickl GmbH, DE) – as the detector of laser-emitted and skin-remitted optical pulses. Time resolution of the system was

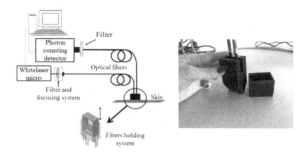

FIGURE 5.2 The skin-remitted laser pulse measurement set-up scheme (left) and a photo of the sliding dual-fiber contact probe (right) [36].

9.7 ps which ensured about 2 mm minimum detectable photon path length in the skin. Five narrow spectral bands centered at 560, 600, 640, 680, 720, and 800 nm were selected by couples of identical interference filters. One of them was filtering the skin input light while the other was placed in front of the photodetector. Stable recording of optical pulsed signals via the input and output fibers (WF-400 from Light Guide Optics Int., LV) was ensured by means of a custom-made fiber holding probe with changeable inter-fiber distances of 1, 8, 12, 16, or 20 mm. To provide equal pressure on the skin surface at all measurements, the probe was designed as a lift where the inside sliding part with a couple of fibers laid on the skin provides a constant pressure of ~35 g/cm² determined by its weight. The outside part of the probe was fixed on the skin during the measurements.

Processing of the measured data involved comparing the normalized shapes of skin input and output pulses – $a(t)$ and $b(t)$, respectively. The temporal distribution function $f(t)$ of photon arrivals following infinitely narrow δ-pulse input was found by de-convolution of the integral $b(t) = \int_{0}^{t} a(t - \tau) f(\tau) d\tau$.

After restoring $f(t)$, the corresponding distribution of back-scattered photon path lengths in the skin was calculated as $\phi(s) = f(t) . c/n$, where c is the speed of light in vacuum and n ~1.4

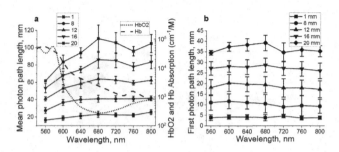

FIGURE 5.3 Spectral dependencies of the mean pathlength of skin-remitted photons at various inter-fiber distances (a) and those calculated for the first detected photons (b) [36]. The dotted curves represent absorption of oxy- and deoxy-hemoglobin [7].

is the mean refraction index of superficial skin tissues. The mean path lengths of skin-remitted photons were found as the mean values of integrated path length distribution functions.

The main obtained results are presented in Figure 5.3. They generally agree with the theoretical predictions that the tissue-scattered photon path length increases with the light input-output distance and with wavelength within the considered spectral range. The obtained numerical values are somewhat higher than those previously predicted by modeling; possible explanations for such discrepancies are provided in [36]. The experimental photon path length data were used for calculations of skin chromophore distribution maps in our further studies; for the wavelength range shorter than 560 nm, the photon path length data were extrapolated from the data measured at 560–800 nm.

The triple spectral line image processing scheme is shown in Figure 5.4. The RGB image taken under three laser line illumination is split into three images – one for each exploited wavelength, applying the RGB crosstalk correction algorithm. Next, the images are segmented to separate the healthy skin from the skin malformation. From the segmented healthy skin, the average signal values at each of the illumination wavelengths (I_0) are

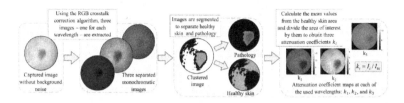

FIGURE 5.4 Three spectral line image processing scheme for obtaining three chromophore distribution maps [37].

calculated; they are used as reference values when there is no additional absorption in the skin. The spectral images (I) are then divided by the reference values, yielding three attenuation coefficient maps $(k(\lambda) = I / I_0)$. These coefficient maps, according to equation (5.2), are further transformed into the distribution maps of chromophore concentration changes in the examined skin malformations.

Proper processing of the three spectral line image ensures the mapping of content changes of the three selected chromophores. In general, the number of chromophores to be mapped is equal to that of the illumination spectral lines used for imaging – it follows from the system of expressions (equation 5.1) with solutions (equation 5.2). For mapping of four chromophore distributions from the data of four spectral line images, a system of four similar expressions must be solved; for a system of five expressions, mapping of five chromophore distributions from the data of five spectral line images can be done, and so on.

5.2 MAPPING OF CHROMOPHORE CONTENT VARIATIONS IN SKIN MALFORMATIONS: CLINICAL DATA

In study [19], vascular and pigmented skin lesions were examined using the SMSLI prototype described in Paragraph 4.2. The examined pathologies included pigmented nevi, seborrheic keratoses, and hemangiomas; all mentioned skin malformations were diagnosed by an experienced dermatologist. The main goal of clinical measurements was to check if the above-described hardware and software can provide physiologically feasible information on already diagnosed skin malformations in terms of increased-decreased contents of the three regarded skin chromophores. Statistical quantitative analysis of skin malformations at this stage of development was not planned.

After processing the clinical images, skin chromophore maps were constructed and changes of malformation's chromophore content with respect to the adjacent healthy skin could be evaluated.

Clinical models assume that the pigmented skin lesions represent an abnormal increase of epidermal melanin and the vascular lesions – increased supply of dermal arterial blood (rich in oxy-hemoglobin) to the superficial layers of skin.

As initially expected, we observed a notable melanin content increase in all cases of pigmented malformations – nevi. An increase in oxy-hemoglobin content and a decrease in deoxy-hemoglobin were observed in all examined hemangiomas (vascular malformations). For illustration, images for three typical clinical cases related to two types of the examined pathologies are presented in Figures 5.5 and 5.6. The color photo of a malformation, taken under simultaneous three wavelength illumination (on the left – a), is compared with the obtained concentration distribution maps of oxy-hemoglobin (b), deoxy-hemoglobin (c), and melanin (d). All pathology data are related to the adjacent healthy skin ($\Delta c = 0$) so that only increased and decreased concentrations are quantified.

Figure 5.5 shows how melanin content increases in nevi, without essential changes in the hemoglobin content. Similar responses were obtained from the seborrheic keratosis. Quite different chromophore composition changes were recorded for vascular hemangiomas (Figure 5.6) – melanin concentration in the pathology remained practically unchanged while the oxy-hemoglobin content notably increased, and the deoxy-hemoglobin content decreased in comparison to healthy skin.

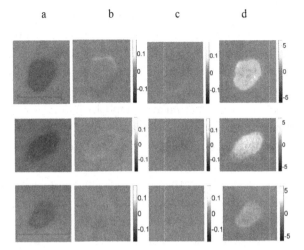

FIGURE 5.5 RGB image (a, scale bar 5 mm) and the corresponding maps representing chromophore concentration changes for three cases of pigmented nevi: b – oxy-hemoglobin, c – deoxy-hemoglobin, d – melanin. Units of the scale on the right – millimoles [19].

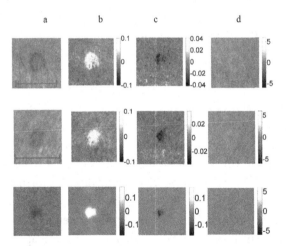

FIGURE 5.6 RGB image (a, scale bar 5 mm) and the corresponding maps representing chromophore concentration changes for three cases of vascular hemangioma: b – oxy-hemoglobin, c – deoxy-hemoglobin, d – melanin. Units of the scale on the right – millimoles [19].

The color scales (on the right) in both figures are calibrated in millimoles (mM), so the increased or decreased chromophore content can be quantitatively analyzed and/or compared. It might help dermatologists to characterize the severity of pathology if clinical thresholds for concentration changes of particular chromophores in the malformation are established.

Skin chromophore distribution can be mapped also by analysis of the conventional, relatively broadband (tens of nm) multispectral image sets – see [8] as an example. A common approach in such cases is to use only the peak wavelength of each spectral band in the calculations of chromophore distributions, ignoring the possibility that the chromophore absorption within the imaging spectral band may change and the mean absorption over the spectral band could differ from that at the band peak wavelength.

SMSLI technique offers a unique opportunity to find out if and how such differences can take place. A study where the same skin malformations were analyzed both under spectral band and spectral line illumination was carried out [38]. Chromophore maps obtained using a ring of LEDs with spectral peaks at 460, 535, and 663 nm for sequential skin illumination were compared with those obtained using the triple laser line illumination device described in Paragraph 4.2. The illumination spectra at both modalities are presented in Figure 5.7, along with the absorption spectra of the three chromophores to be mapped. In all cases, RGB camera of the smartphone Nexus5 was used to capture the spectral band images and spectral line images of the same skin malformations.

A significant number of pigmented formations (19 junctional nevi, 23 dermal nevi, 19 combined nevi), vascular formations (21 hemangiomas), as well as 23 seborrheic keratoses were analyzed in this study. The obtained results demonstrated essential differences in the chromophore maps calculated from images taken at spectral band and spectral line illumination – see Figure 5.8 as an example. Moreover, quite notable differences were observed also regarding the concentration changes of three chromophores in the examined pathology groups, as presented in Figure 5.9. For instance, oxy-hemoglobin content in hemangiomas does not

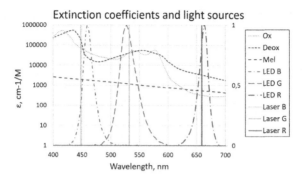

FIGURE 5.7 Normalized spectra of the LED and laser illuminators and absorption spectra of three main skin chromophores: Ox, oxy-hemoglobin; Deox, deoxy-hemoglobin; Mel, melanin; LED B, blue LEDs; LED G, green LEDs; LED R, red LEDs; Laser B, blue laser; Laser G, green laser; Laser R, red laser [38].

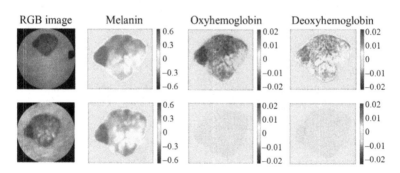

FIGURE 5.8 Example of RGB images of a seborrheic keratosis and the corresponding chromophore maps of the content changes for melanin, oxy-hemoglobin and deoxyhemoglobin, calculated from the images taken at spectral band (upper row) and spectral line (lower raw) illumination [38].

differ much from the other considered skin malformations if the spectral band images are processed (Figure 5.9c). The spectral line image processing, in opposite, clearly distinguishes hemangiomas (Figure 5.9d). Also, the other examined pathologies are better distinguished if the triple spectral line imaging technology is applied.

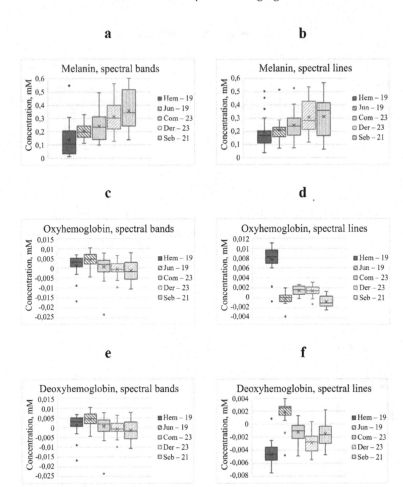

FIGURE 5.9 Comparison of the relative concentration changes of melanin (a, b), oxyhemoglobin (c, d), and deoxyhemoglobin (e, f) in hemangiomas (*Hem*), junctional nevi (*Jun*), combined nevi (*Com*), dermal nevi (*Der*), and seborrheic keratoses (*Seb*), calculated for the cases of the spectral band (a, c, e) and the spectral line (b, d, f) illumination. Rectangles in graphs describe 25%–75% values, vertical lines – standard deviations, crosses – mean values, horizontal lines – median values, circles – outliers [38].

Generally, the provided comparison confirms that the chromo-phore map calculations from spectral band images, considering only the peak wavelengths of imaging bands, lead to questionable results.

Obviously, image processing in such cases should involve integration over all wavelengths comprised within each spectral band of imaging, which is a much more complicated procedure. Anyway, clear advantages of the spectral line imaging approach for skin chromophore 2D-mapping have been demonstrated by this study.

The SMLI technology also offers other new challenges for non-contact skin diagnostics. For example, 3D-representation of the triple spectral line imaging data for improved diagnostics was proposed [37,39]. In this study, clinical in vivo measurements were taken of 77 white-skin volunteers in cooperation with a certified dermatologist. In total 99 skin pathologies were examined: 3 basal cell carcinomas, 27 dermal nevi, 12 hemangiomas, 16 combined nevi, 1 melanoma, 17 junctional nevi, 22 seborrheic keratoses, and 1 blue nevi. Data for dermal, combined, and junctional nevi, as well as those for hemangiomas and seborrheic keratoses may be considered as sufficient for primary conclusions, while the few cases of skin cancers (basal cell carcinoma, melanoma) and blue nevi can serve only for general illustration.

The triple laser line prototype (Paragraph 4.2) was used to measure the reflected intensity ratios – pathology vs healthy skin, further named as attenuation coefficients k_i – at the three working wavelengths for the above-mentioned eight groups of pathologies. The k-values related to 448 nm are denoted as k_B, those related to 532 nm as k_G, and those related to 659 nm as k_R. Each of them represents one of the three coordinate axes in the 3D-graphs shown in Figure 5.10. Every point in the graph rep-resents one clinical data pixel signal value from a segmented pathology. Points are arranged more densely where pixel output

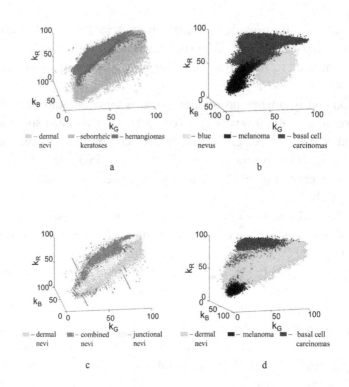

FIGURE 5.10   3D graphs of the three attenuation coefficients (in %) for different groups of skin pathologies: dermal nevus, seborrheic keratosis, and hemangioma (a); blue nevus, melanoma, and basal cell carcinoma (b); dermal, combined and junctional nevus (c); dermal nevus, melanoma, and basal cell carcinoma (d) [37].

signals have similar values, and they are more scattered where only a few pixels have these values. Each cloud consists of 6,000–400,000 non-overlapping points.

The k-value clouds for three different benign pathologies – dermal nevi, seborrheic keratoses, and hemangiomas – are compared in Figure 5.10a. Parts of them are non-overlapping, e.g., the specific volume related to hemangiomas and that related to nevi can be easily distinguished. Even better separation between malformations can be observed in Figure 5.10b where spectral attenuations of two malignant pathologies – melanoma and basal cell carcinomas – are compared with those of a benign pathology – blue nevus. All three pathologies here can easily be distinguished; the k_R values for blue nevus and melanoma are lower than those for basal cell carcinoma while melanoma exhibits lower k_G values than the blue nevus. Spectral attenuations of three different types of nevi – dermal, combined, and junctional – are compared in Figure 5.10c; they form a compact cloud but still each type mainly covers a specific volume in the 3D graph. Data for malignant pathologies (basal cell carcinoma and melanoma) and typical benign pathology (dermal nevus) are compared in Figure 5.10d. Again, some values of all three malformations are slightly overlapping but the k_R values for nevi are clearly higher than those for melanoma and lower than those for basal cell carcinomas. To summarize, the obtained 3D-graphs exhibit specific volume-shape features for each of the examined eight groups of skin pathologies. This kind of spectral line image data representation may find further application in quantitative diagnostics of skin malformations, e.g., for characterizing and identifying specific skin pathologies by the spatial location of the $k_B - k_G - k_R$ data points determined from the spectral line images.

Besides, such three-dimensional spectral attenuation data set can be further transformed into 3D-maps representing the content increase or decrease of the main skin chromophores in the regarded malformations. Figure 5.11 shows the results of calculated chromophore concentration changes which are represented in colors according to the calibrated color scales on the right from each "cloud." Only points representing the pixel signal values in clinical data images were left in the 3D-graphs of spectral attenuation coefficients. Green color in all graphs represents zero changes in chromophore concentrations. If relative chromophore content is higher in the pathology than in the adjacent healthy skin, the points in the graph are colored yellow or red; if the chromophore content has decreased, the points are colored blue. The melanin, oxyhemoglobin, and deoxyhemoglobin relative concentration scales are leveled equally for all included pathologies.

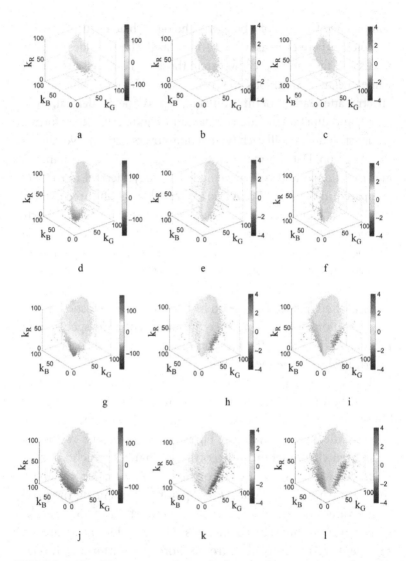

a

b

c

d

e

f

g

h

i

j

k

l

FIGURE 5.11 3D-representation of chromophore content changes in four examined skin pathologies. The chromophore concentration increase/decrease in the pathologies is graduated in moles·10^{-5} [37].

Melanin Oxy-hemoglobin Deoxy-hemoglobin

One can notice that the spatial cloud shapes are specifically different for each of the examined eight types of skin pathologies. Melanomas and nevi exhibit the highest increase in melanin concentration, whereas the oxyhemoglobin concentration is highest for hemangiomas, as could be expected from the lesion anatomy considerations. Although some of the obtained 3D clouds for particular skin malformations spatially overlap, individual features of specific pathology groups like cloud shape/volume and chromophore content distribution could be clearly distinguished. The spectral imaging data were processed in the frame of a simple BLL model where only relative concentration changes of three main skin chromophores (melanin, oxyhemoglobin, and deoxyhemoglobin) with respect to the surrounding healthy skin were calculated. Despite evident limitations of the used model, the proposed 3D-representation approach demonstrated not only selectivity concerning the type of pathology but also some advantages if compared to the commonly used approach – planar or 2D-mapping of the chromophore distributions over the skin malformation. The planar chromophore maps are well-suited for assessment of a specific single pathology sample but this representation lacks sense of general trends of chromophore distributions in the pathology groups.

Qualitatively, all hemangiomas exhibited increased oxyhemoglobin concentration than other examined pathologies, and all pigmented malformations (nevi, basaliomas, melanoma) exhibited increased melanin concentration. The 3D representation of spectral imaging data potentially enables objective comparison of different pathology groups and may facilitate a general understanding of the chromophore content variations and their distributions in skin malformations. Further development of this approach, e.g., by applying AI algorithms for 3D cloud recognition, could be helpful for healing follow-up and for improved quantitative diagnostics of particular pathologies, including skin cancers.

Clinical validation of the snapshot four spectral line imaging prototype (Paragraph 4.4) has been initiated, as well [14]. This device enables mapping four skin chromophores instead of three, so skin lipids were chosen as the new/additional chromophore. Depending on the wavelength, the contrast in the spectral line images changed due to the different penetration depths in the skin. In the case of dermal nevus – a pigmented lesion located below the epidermal/dermal junction – notably increased melanin content was observed with slight changes in the concentrations of three other chromophores. Hemangioma is a vascular malformation where an increased content of oxy-hemoglobin (in relation to the surrounding healthy skin) was observed, accompanied with a decreased content of deoxy-hemoglobin. Seborrheic keratosis, frequently misdiagnosed as malignant melanoma, also exhibited increased melanin content.

5.3 WHOLE BODY SPECTRAL LINE IMAGING: CLINICAL DATA

The whole-body spectral line imaging prototype (Paragraph 4.6) passed preliminary clinical validation [31]. The protocol, approved by the local ethics committee, comprised six fixed body positions of the patient; the imaging procedure typically took less than 2 minutes. The image processing software included the detection of all notable (1 mm or larger) skin lesions, highlighting the healthy skin regions in L*a*b color scale with subsequent calculation of the binary mask, identification and numbering of the lesions, and composing the album of detected lesion's images (Figure 5.12). Each of the album's images was further spectrally processed to find out the type of lesion – pigmented or vascular. If pigmented, the melanin increase level was estimated and

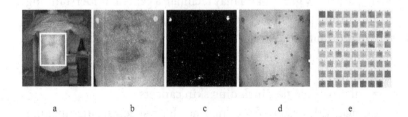

a   b   c   d   e

FIGURE 5.12   The large area skin image processing procedure: a – initial RGB image, b – highlighted healthy skin, c – the calculated binary mask, d – identification and numbering of the lesions, e – album of the detected lesion's images [31].

compared to the melanoma risk threshold, evaluated according to the findings of [40,41]. Extracting three spectral line images at 450, 520, and 638 nm from the obtained RGB image data and mapping the content increase/decrease of three main skin chromophores over the examined malformations could be performed by a technique described above.

As a clinical example, Figure 5.13 (middle) presents a high-resolution image of volunteer's back taken under triple spectral line illumination. A malformation sized ~1 mm is selected (see the arrow) and further analyzed. The three extracted spectral line images of this malformation are presented on the right side, while the distribution maps representing content changes of three main skin chromophores (melanin, oxy-hemoglobin, and deoxy-hemoglobin), calculated from the spectral line images, are presented on the left side.

Considering that the melanin content in this malformation has notably increased relatively to the surrounding healthy skin and practically no changes in the content of oxy-hemoglobin

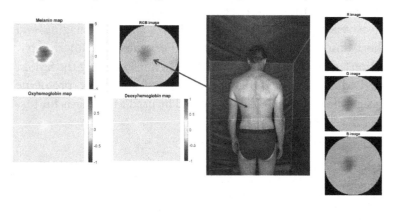

a   b   c

FIGURE 5.13 The calculated maps of three skin chromophore content changes (a) of a small nevus on the volunteer's back (b), and three extracted spectral line images of this nevus (c), where R – 638 nm, G – 520 nm, B – 450 nm. Color scale (a) in millimoles [31].

and deoxy-hemoglobin can be observed, this lesion is identified as a pigmented nevus. The sorting criterium for vascular lesions is increased content of oxy-hemoglobin in conjunction with decreased content of deoxy-hemoglobin and minimal or no changes in the melanin content [19].

To conclude, this chapter demonstrated the clinical utility of the lab-developed multi-spectral line imaging prototypes. SMSLI technology proved to be well-suited for advanced 2D and 3D mapping of the three main chromophore content changes in various skin malformations. The increase/decrease of chromophore concentrations could be quantified (e.g., in millimoles), which allows objective diagnostics and comparison with the threshold values of the pathology's severity. The recently started whole-body triple spectral line imaging seems to be a promising technique for fast remote multi-lesion analysis and, eventually, early screening of skin cancer. Besides, the SMSLI technology may be adapted for diagnostic and follow-up applications in other clinical fields including endoscopy, ophthalmology, dentistry, cosmetology, and surgery. The main expected advantages are all-in-one non-invasiveness, high spectral and spatial resolution of imaging, and short time of the procedure. To facilitate further clinical applications, new clinically certified devices implementing the SMSLI method should be developed; the proof-of-concept prototype designs described in Chapter 4 could be used as the basis for further developments.

Forensic Applications

A ROUND A MILLION COUNTERFEIT money banknotes are withdrawn from worldwide circulation annually. Even assuming that all fake banknotes have been detected, the losses of global economy due to forgeries reach hundreds of millions USD. To be realistic, a large number of fake banknotes still circulate as the criminal world rapidly adapts to the existing technologies of counterfeit detection, so high-quality fake banknotes frequently are recognized as genuine by the serial forgery detectors. In the case of Euros, it especially relates to professionally printed €500 banknotes [42], therefore new €500 banknotes are not issued any more but those produced previously still circulate.

To reduce the huge losses for economy and citizens, new non-conventional forgery detection technologies should be continuously developed and implemented. Novel techniques are also demanded to discover increasing forgeries of personal documents – e.g., ID cards, visas, and other travel documents used for illegal immigration and other criminal activities [43]. Signature forgeries on legal documents with amended content or replaced pages are also widely used [44], as well as forgeries of paintings, design items, brand labels, and other colored objects.

DOI: 10.1201/9781003476702-7

For obvious reasons, the technologies for forgery detection and the related device designs are not widely advertised. In general, the main items to examine on banknotes and documents are the texture, watermarks and other paper features, the hidden infrared, UV-fluorescent and/or magnetic markers, the used printing/copying technique, specific pattern details, and their colors. Along with visual determination/comparison of the colors, the spectral distribution of light reflected from the examined banknote or document area is an important source of information. Two objects comprising different combinations of pigments (each of them with its own specific absorption spectra) visually may appear of the same color in white light but look entirely different under spectrally narrowband illumination, as one pigment can absorb stronger and some other pigment(s) weaker within the selected spectral band. Comparison of spectral images taken from the examined sample (e.g., banknote) and from a genuine reference standard allows distinguishing between the original ink, toner, or paint and those used for forgeries.

Clearly, the narrower is the spectral band of imaging, the pigment discrimination sensitivity is higher – from this point, the multi-spectral-line imaging technology seems very promising. This expectation was confirmed by the results of our initial study, presented in this chapter.

6.1 COUNTERFEIT BANKNOTE DETECTION

To assess the potential of SMSLI technology for the detection of colored counterfeits, the developed triple-wavelength device (Paragraph 4.2) was used for comparative analysis of genuine and counterfeit banknotes in collaboration with the Bank of Latvia. The very first results with €20 banknotes were promising. We found that the most sensitive among the three exploited imaging wavelengths was 448 nm (Figure 6.1), but even more differences between the genuine and counterfeit banknote appeared if the ratios of spectral line images involving the blue laser line image were calculated. Figure 6.2 convincingly demonstrates

a   b   c

FIGURE 6.1 Specimen of €20 banknote with marked region of interest (RoI) (a) and its 448 nm spectral line images for the authentic (b) and counterfeit (c) banknote [26].

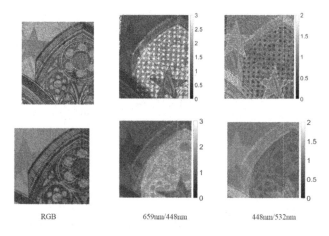

RGB   659nm/448nm   448nm/532nm

FIGURE 6.2 Comparison of RGB images and spectral line image ratios for the same fragment of authentic (upper raw) and counterfeit (lower raw) €20 banknotes [26].

how the ratios of spectral line images for the same fragment of €20 banknote clearly distinguish the fake banknote from the genuine one, even if visually (RGB images on the left) they look very similar.

This result facilitated starting more systematic studies. An algorithm for processing of the banknote spectral line images was developed including a quantitative measure – k-factor, representing the contrast of the selected element relatively to its

Banknote image processing

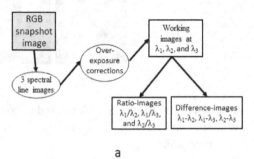

a

k-factor of the selected pixel group

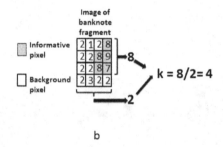

b

FIGURE 6.3 Processing scheme of the banknote spectral line images (a) and illustration for k-factor calculation of a selected pixel group (b) [26,45,46].

background, as presented in Figure 6.3. For each pixel group (related to a specific spot on the banknote), the k-factor was calculated as the ratio of mean signal values of the informative and the background pixels. If necessary, over-exposure corrections were performed after the spectral line image acquisition, and then three working images of the same spot were mutually divided and/or extracted. The comparison of k-values further served for distinguishing between authentic and false banknotes.

Several kinds of counterfeit methods were considered – jet-printed, offset-printed, color copied, and some unclassified

(a)

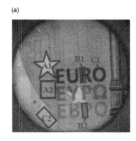

(b)

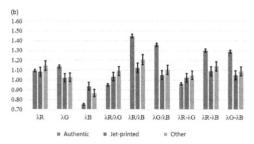

FIGURE 6.4   The elements selected for examination on the 50 EUR banknotes (a) and the k-factor values (A1/A2) for authentic banknotes and two groups of counterfeits – jet-printed and other [45].

methods. In study [45], altogether 58 authentic and counterfeit €20 and €50 banknotes were examined. Counterfeit banknotes were initially classified according to the production technology identified by the Money Technology Department of the Bank of Latvia. Six different elements on €50 banknotes were analyzed (Figure 6.4a). A1 element is one of the stars on €50 banknotes front side, while A2 is the background near A1. Star element was chosen because one of its components is europium oxide with luminescent properties. The k-factor – the ratio of mean values of the informative (the star) and the background pixels (A1/A2) – was calculated for three spectral line images, one for each of the used wavelength: 448 nm (λ_B), 532 nm (λ_G), and 659 nm (λ_R), along with all spectral image ratios and their differences (Figure 6.4c). One can see that the best original/counterfeit separation is provided by the blue 448 nm spectral line image, as well as by all its ratios and differences with respect to the green and red line images.

Besides, spectral line images of genuine and fake €500 banknotes were analyzed [47]. This nominal is usually forged with high quality, so only professionally counterfeited banknotes were examined. Three spots of the banknotes were analyzed to identify the most vulnerable locations where forgery could be detected by the spectral line image analysis (Figure 6.5).

FIGURE 6.5 The specimen of 500 EUR banknote from the front (a) and back (b) with outlined three regions of interest [47].

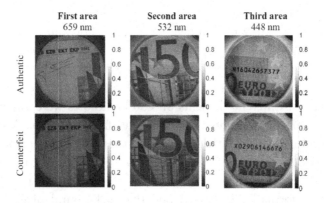

FIGURE 6.6 Spectral line images of authentic and professionally forged 500 EUR banknotes at the three selected areas of interest, each under a different illumination wavelength [47].

Comparison of the spectral line images for the €500 banknotes is presented in Figure 6.6. In the first selected area, the red 659 nm spectral line images comprising the signature and stars are compared. Visually, the stars look the same on both banknotes; under 659 nm illumination, the star on the right side of the authentic banknote absorbs less light than the background around it, while the star on the counterfeit appears darker. Besides, the pattern in the bottom right corner looks different. In the second selected area, buildings under the number "5" were examined using the green 532 nm spectral line images. In the counterfeit banknote, several parallel stripes over the building are visible, which are absent in the authentic banknote. The stripes may be caused by the printer used for forgery. In the third area under the blue 448 nm line illumination, the rectangular corner is missing in the fake banknote. Besides, the words "EURO" and "EYPΩ" are well contrasted in the authentic banknote, while they blend into the background in the case of counterfeit banknote.

6.2 DETECTION OF DOCUMENT COUNTERFEITS

An international study was initiated to compare various technologies, including multi-spectral-line imaging, for the detection of counterfeits in documents [48]. Potentially fake documents consisting of three pages were created in the forensic laboratory of the Portuguese police. Identical samples, simulating a suspected intentional manipulation of a real estate rental agreement consisting of three pages, were sent to 17 laboratories in 16 different countries. The aim was to determine which printing technique was used, whether the same printer was used to print all three pages, whether all the pages are of the same paper, whether they were all originally stapled together, etc. The second page in the sample document was printed by a different printer on a different type of paper in comparison to the first and third pages.

Each of the three pages of the sample document was individually examined using the triple laser line prototype described in Paragraph 4.2. Average black toner pixel values for printed texts were calculated at each of the used wavelength; they are presented in Table 6.1. According to the obtained results, the calculated average values for the first and third pages are very similar, but they are significantly different for the second page. This indicates that a different printer or ink cartridge was used to print the second page.

To conclude, forensics appears to be one of the promising future applications of SMSLI technology. Our initial studies confirmed high sensitivity of counterfeit detection related to EUR banknotes and printed documents. The potential applications

TABLE 6.1 Three Spectral Line Images and the Calculated Average Pixel Values for Black-Printed Texts Related to Three Pages of a Document [47,48]

Nr.	659 nm	Av. Value, $\cdot 10^{-2}$	532 nm	Av. Value, $\cdot 10^{-2}$	448 nm	Av. Value, $\cdot 10^{-2}$
1	ntos para / na rua Fl / de Sant / sob o ar	2.9 ± 1.5	ntos para / na rua Fl / de Sant / sob o ar	7.8 ± 3.5	ntos para / na rua Fl / de Sant / sob o ar	5.1 ± 0.4
2	nforme / nalizaçõe / ção, toma / ta as nec	4.2 ± 1.7	nforme / nalizaçõe / ção, toma / ta as nec	9.8 ± 3.6	nforme / nalizaçõe / ção, toma / ta as nec	6.3 ± 0.4
3	rgante c / umo de / trica que	2.8 ± 1.5	rgante c / umo de / trica que	7.3 ± 3.3	rgante c / umo de / trica que	5.1 ± 0.4

can be extended to other currencies, including USD, as well as to personal documents like ID-cards, driver's licenses, and visa permissions. The exceptionally high color sensitivity of the method could contribute to better recognition of false paintings and other artworks, as well. Surprisingly, the ultra-narrowband multi-spectral-line imaging demonstrated high sensitivity not only to color pigment compositions but also to different black pigments. This confirms the expectations to see even more fascinating forensic applications of the SMSLI technology in the future.

Summary

"Take-Home Messages"

1. The spectral content of illumination and spectral sensitivity of the image sensor play an important role in digital camera-based photography; if optimized, they allow extending the conventional RGB color imaging to high-performance multispectral imaging.

2. Improving multispectral imaging technique by narrowing the spectral bands of imaging down to spectral linewidths and simultaneous capturing of several narrowband spectral images are proposed using the snapshot multi-spectral-line imaging (SMSLI) method, described in Chapter 2.

3. To implement SMSLI, specific conditions for image sensor and illumination should be met:

   a. Only image sensors with linear photo-response and known relative photo-sensitivities at the working wavelengths are suitable.

   b. Uniform illumination over the target area simultaneously by several discrete spectral lines must be ensured; multilaser illumination is an option.

4. If applying lasers, their narrow-collimated output beams should be expanded and/or scattered; four such illumination designs are presented in Chapter 3, along with the measures to fight laser speckle artifacts in spectral images.

5. The viability of the SMSLI concept and the proposed illumination designs have been practically proven by ten lab-made prototype devices (Chapter 4), validated in laboratory and clinical environments.

6. The application potential of SMSLI technology for improved clinical diagnostics and highly sensitive forgery detection was confirmed by the measurement data presented in Chapters 5 and 6, respectively.

7. The high spectral and tempo-spatial selectivity offered by SMSLI may facilitate future developments in healthcare (e.g., dermatology, cosmetology, endoscopy, dentistry), forensics (detection of fake banknotes, paintings, personal and legal documents), and other application areas of multispectral imaging.

Acknowledgments

THE CONTENT OF THIS book is based on ideas and devices created in collaboration with colleagues and students to whom I am sincerely grateful for their contributions. Starting with the introduction of MSI principles to my Master's student (now PhD) Dainis Jakovels by Professor Goran Salerud in Linkoping University back in 2010, this "multispectral infection" gradually spread out in our laboratory. Dainis assembled his first LED-based portable prototype for triple spectral band imaging and initiated studies of skin hemoglobin mapping under dual laser line illumination, which laid the fundament for further developments of the SMSLI technology. After assembling the first laboratory prototype (Figure 4.1), BSc students Liene Elste and Jekabs Zarins completed the first proof-of-concept measurements with this device in 2012–2014. Later our lab colleagues Dr. Edgars Kviesis-Kipge and Dr. Uldis Rubins together with another team of students – Ilze Oshina (now PhD), Zigmars Rupenheits, Kalvis Lauberts – contributed to the development of several more advanced SMSLI designs (Figures 4.2–4.8). Ilze started her PhD studies on this topic and actively participated in the clinical validation trials together with MD Anna Berzina and Dr. Ilona Kuzmina, as well as in the forgery studies together with BSc student Peteris Potapovs. Madars Mileiko significantly contributed to mechanical designs during the period 2020–2023. All the abovementioned people essentially promoted developments of the SMSLI technology.

I had the privilege to lead several research projects carried out on this topic in our lab, in particular:

"Smart Materials, Photonics, Technologies and Engineering Ecosystem", Latvian Council of Science project # VPP-EM-FOTONIKA-2022/1-0001, 2022-2024.

"Multimodal imaging technology for in-vivo diagnostics of skin malformations", European Regional Development Fund project # 1.1.1.1/18/A/132, 2019-2022.

"Advanced spectral imaging technology for skin diagnostics", Latvian Council of Science project # lzp-2018/2-0006, 2018-2020.

"Biophotonics: imaging, diagnostics and monitoring" – project #3 of Latvian National Research Program "SOPHIS" #10-4/VPP-4/11, 2014-2017.

"Innovative technologies for diagnostic imaging of skin" – European Regional Development Fund project #2111/3 -015, 2014-2015.

"New optical technologies for complex non-contact skin diagnostics" – European Regional Development Fund project #2111-030, 2010-2013.

"Biophotonics research group", European Social Fund project #1112-077, 2009-2012.

My sincere thanks to the editors of this book for their valuable comments and corrections. At last (but not least), many thanks to my family for support during the writing period.

References

1. X. Jia (2022). *Field Guide to Hyperspectral / Multispectral Image Processing*, Vol. FG52, SPIE Press, Bellingham, WA, 118 p (2022).
2. C.-I Chang, M. Song, J. Zhang, eds. (2022). *Hyperspectral Imaging and Applications*. MDPI books, Basel, CH, 632 p (2022).
3. Y. Garini, E. Tauber (2013). "Spectral imaging: Methods, design, and applications", In: Liang, R. (ed), *Biomedical Optical Imaging Technologies. Biological and Medical Physics, Biomedical Engineering*. Springer, Berlin, Heidelberg, p.111–158 (2013).
4. M.A. Ilişanu, F. Moldoveanu, A. Moldoveanu, "Multispectral imaging for skin diseases assessment - State of the art and perspectives", *Sensors* 23(8), 3888 (2023). DOI: 10.3390/s23083888.
5. L. Huang, R. Luo, X. Liu, et al., "Spectral imaging with deep learning", *Light Sci. Appl.* 11(1), 61 (2022). DOI: 10.1038/s41377-022-00743-6.
6. RP Photonics Encyclopedia, https://www.rp-photonics.com/narrow_linewidth_lasers.html. Last access 23.02.2024.
7. S. Prahl, Tabulated molar extinction coefficient for hemoglobin in water, https://omlc.ogi.edu/spectra/hemoglobin/summary.html. Last access 23.02.2024.
8. D. Jakovels, J. Spigulis, "2-D mapping of skin chromophores in the spectral range 500-700 nm", *J. Biophoton.* 3(3), 125–129 (2010). DOI: 10.1002/jbio.200910069.
9. D. Jakovels, J. Spigulis, "Skin haemoglobin mapping: Comparison of multi-spectral imaging and selective R-G-B analysis", In: *Biomedical Optics and 3-D Imaging*, OSA Technical Digest (CD), Optica Publishing Group, Washington, USA (2010), paper BTuD62. https://opg.optica.org/abstract.cfm?URI=BIOMED-2010-BTuD62
10. J. Spigulis, D. Jakovels, L. Elste, "Towards single snapshot multispectral skin assessment", *Proc. SPIE* 8216, 82160L (2012). DOI: 10.1117/12.908967.
11. J. Spigulis, L. Elste, "Single-snapshot RGB multispectral imaging at fixed wavelengths: Proof of concept", *Proc. SPIE* 8937, 89370L (2014). DOI: 10.1117/12.2039442

12. J. Spigulis, L. Elste, "Method and device for imaging of spectral reflectance at several wavelength bands", Patent *WO2013135311 (A1)* (2012). https://smallbusiness.chron.com/wo-patent-64013.html
13. J.Spigulis, "Method for imaging of spectral reflectance at several wavelengths", Patent WO2015071786 (A1) (2015).
14. J. Spigulis, Z. Rupenheits, U. Rubins, M. Mileiko, I. Oshina, "Spectral line reflectance and fluorescence imaging device for skin diagnostics", *Appl. Sci.* 10, 7472 (2020). DOI: 10.3390/app10217472.
15. J. Spigulis, I. Oshina, "Method and device for mapping of chromophores under several spectral line illumination", patent LV 15106 B (2016).
16. J. Spigulis, I. Oshina, I. Kuzmina, L. Dambite, "Method and device for determination of photo-camera relative spectral sensitivity at selected wavelengths", patent LV 15705 B (2023).
17. J. Spigulis, I. Oshina, "Snapshot RGB mapping of skin melanin and hemoglobin", *J. Biomed. Opt.* 20(5), 050503 (2015). DOI: 10.1117/1.JBO. 20.5.050503.
18. J. Spigulis, M. Lacis, I. Kuzmina, A. Lihacovs, V. Upmalis, Z. Rupenheits, "Method and device for smartphone mapping of tissue compounds", patent *WO 2017/012675 A1* (2017).
19. J. Spigulis, I. Oshina, A. Berzina, A. Bykov, "Smartphone snapshot mapping of skin chromophores under triple-wavelength laser illumination", *J. Biomed. Opt.* 22(9), 091508 (2017). DOI: 10.1117/1.JBO.22.9.091508.
20. D. Pfafrods, M. Stafeckis, J. Spigulis, D. Boucher, "Side-emitting optical fiber", patent LV 11644 B (1995).
21. J. Spigulis, D. Pfafrods, M. Stafeckis, W. Jelinska-Platace, "The "glowing" optical fibre designs and parameters", *Proc. SPIE* 2967, 226–231 (1997). DOI: 10.1117/12.266542.
22. J. Spigulis, "Side-emitting optical fibers brighten our world in new ways", *Opt. Photonics News* 16(10), 34–39 (2005). DOI: 10.1364/OPN.16. 10.000034.
23. J. Spigulis, I. Oshina, Z. Rupenheits, M. Matulenko, "Device for uniform illumination simultaneously by several laser spectral lines", patent LV 15491 B (2020).
24. U. Rubins, E. Kviesis-Kipge, J. Spigulis, "Device for speckle-free imaging under laser illumination", patent *WO 2018/177565 A1* (2018).
25. J. Spigulis, Z. Rupenheits, M. Matulenko, I. Oshina, U. Rubins, "A snapshot multi-wavelengths imaging device for *in-vivo* skin diagnostics", *Proc. SPIE* 11232, 112320I–1 (2020). DOI: 10.1117/12.2547286.
26. J.Spigulis, I.Oshina, P. Potapovs, K. Lauberts, "Snapshot multi-spectral-line imaging for applications in dermatology and forensics", *Proc. SPIE* 10881, 1088114 (2019). DOI: 10.1117/12.2508204.
27. I. Oshina, J. Spigulis, U. Rubins, E. Kviesis-Kipge, K. Lauberts, "Express RGB mapping of three to five skin chromophores", *Proc. SPIE-OSA* 10413, 104130M–1 (2017). DOI: 10.1117/12.2285995.

28. J. Spigulis, I. Oshina, M. Matulenko, "Laser illumination designs for snapshot multi-spectral-line imaging", *IEEE Xplore* (2019). https://ieeexplore.ieee.org/document/8872998.

29. E. Kviesis-Kipge, "Development of skin chromophore mapping device using five spectral line illumination", *OSA Tech. Dig.* (2019), ITh4B.3, DOI: 10.1364/ISA.2019.ITh4B.3.

30. J. Spigulis, E. Kviesis-Kipge, U. Rubins, I. Oshina, M. Mileiko, "RGB laser-illuminated spectral imaging: Applications in dermatology and endoscopy", *Proc. IFMBE* 89, 138–144 (2023). DOI: 10.1007/978-3-031-37132-5_18.

31. J. Spigulis, U. Rubins, E. Kviesis-Kipge, I. Saknite, I. Oshina, E. Vasilisina, A. K. Krievina, "Diagnostic spectral imaging of skin and nasal mucosa by RGB laser-based prototype devices", *Proc. SPIE* 13009, 130090C (2024). doi: 10.1117/12.3018710.

32. J. Spigulis, I. Oshina, "3x3 technique for RGB snapshot mapping of skin chromophores", In: *Optics in the Life Sciences*, OSA Technical Digest (online), Optical Society of America (2015), paper JT3A.39, https://opg.optica.org/abstract.cfm?uri=BODA-2015-JT3A.39.

33. RP Photonics Encyclopedia, Beer-Lambert law, https://www.rp-photonics.com/beer_lambert_law.html. Last access 03.06.2024.

34. I. Oshina, J. Spigulis, "Beer-Lambert law for optical tissue diagnostics: Current state of the art and the main limitations", *J. Biomed. Opt.* 26(10), 100901 (2021). DOI: 10.1117/1.JBO.26.10.100901.

35. T. Sarna, H. M. Swartz, "The physical properties of melanin", https://omlc.ogi.edu/spectra/melanin/eumelanin.html. Last access 23.02.2023.

36. V. Lukinsone, A. Maslobojeva, U. Rubins, M. Kuzminskis, M. Osis, J. Spigulis, "Remitted photon path lengths in human skin: *in-vivo* measurement data", *Biomed. Opt. Expr.* 11(5), 2866–2873 (2020). DOI: 10.1364/BOE.388349.

37. I. Oshina, J. Spigulis, I. Kuzmina, L. Dambite, A. Berzina, "Three-dimensional representation of triple spectral line imaging data as an option for noncontact skin diagnostics", *J. Biomed. Opt.* 27(9), 095005 (2022). DOI: 10.1117/1.JBO.27.9.095005.

38. I. Kuzmina, I. Oshina, L. Dambite, V. Lukinsone, A. Maslobojeva, A. Berzina, J. Spigulis, "Skin chromophore mapping by smartphone RGB camera under spectral band and spectral line illumination", *J. Biomed. Opt.* 27(2), 026004 (2022). DOI: 10.1117/1.JBO.27.2.026004.

39. I. Oshina, J. Spigulis, "3D-representation of skin malformations using spectral line imaging and modified Beer-Lambert law", *Proc. SPIE* 12665, 126650T (2023). DOI: 10.1117/12.2687959.

40. I. Kuzmina, I. Diebele, D. Jakovels, J. Spigulis, L. Valeine, J. Kapostinsh, "Towards noncontact skin melanoma selection by multispectral imaging analysis", *J. Biomed. Opt.* 16(6), 060502 (2011). DOI: 10.1117/1.3584846.

41. I. Diebele, I. Kuzmina, A. Lihachev, J. Spigulis, J. Kapostinsh, A. Derjabo, L. Valaine, "Clinical evaluation of melanomas and common nevi by spectral imaging", *Biomed. Opt. Express* 3(3), 467–472 (2012). DOI: 10.1364/BOE.3.000467.

42. H. Thompson, "Warning over false (but legal) euro notes used by fraudsters in France", (2022). https://www.connexionfrance.com/article/French-news/Warning-over-false-but-legal-euro-notes-used-by-fraudsters-in-France. Last access 23.02.2024.

43. H. Esteves, "Introduction to fraudulent methods used in travel, identity and visa documents", (2012). https://www.icao.int/meetings/mrtd-zimbabwe2012/documents/2-11-esteves_portugal-forensic.pdf. Last access 23.02.202.

44. Upcouncel, "False signature on contract: Everything you need to know", https://www.upcounsel.com/false-signature-on-contract. Last access 23.02.2024.

45. I. Oshina, P. Potapovs, J. Spigulis, "Spectral imaging system for money counterfeit detection", *OSA Tech. Dig.* ITu3B.3 (2019). DOI: 10.1364/ISA.2019.ITu3B.3.

46. J. Spigulis, I. Oshina, "A method for detection of colored counterfeits", patent LV 15413 (2019).

47. I. Oshina, "Spectral line imaging for non-contact skin diagnostics and counterfeit detection", Summary of PhD Thesis, University of Latvia, Riga (2023).

48. T. Fisher, M. Marchetti-Deshmann, A.C. Assis, et al., "Profiling and imaging of forensic evidence - a pan-European forensic round robin study part 1: Document forgery", *Sci. Justice* 62(4), 433–447 (2022). https://doi.org/10.1016/j.scijus.2022.06.001

Index

Note: **Bold** page numbers refer to tables and *italic* page numbers refer to figures.

Printed in the United States
by Baker & Taylor Publisher Services